# Fourier Transform Mass Spectrometry

ACS SYMPOSIUM SERIES **359**

# Fourier Transform Mass Spectrometry

## Evolution, Innovation, and Applications

**Michelle V. Buchanan**, EDITOR
*Oak Ridge National Laboratory*

Developed from a symposium sponsored
by the Division of Analytical Chemistry
at the 192nd Meeting
of the American Chemical Society,
Anaheim, California,
September 7-12, 1986

American Chemical Society, Washington, DC 1987

**Library of Congress Cataloging-in-Publication Data**

Fourier transform mass spectrometry: evolution, innovation, and applications/Michelle V. Buchanan, editor.

p. cm.—(ACS symposium series, ISSN 0097-6156; 359)

"Developed from a symposium sponsored by the Division of Analytical Chemistry at the 192nd Meeting of the American Chemical Society, Anaheim, California, September 7-12, 1986."

Includes bibliographies and indexes.

ISBN 0-8412-1441-7

1. Ion cyclotron resonance spectroscopy—Congresses. 2. Fourier transform spectroscopy—Congresses.

I. Buchanan, Michelle V., 1951- . II. American Chemical Society. Division of Analytical Chemistry. III. American Chemical Society. Meeting (192nd: 1986: Anaheim, Calif.) IV. Series.

QD96.I54F68 1987
543'.0873'015157—dc19 87-26107
CIP



PRINTED IN THE UNITED STATES OF AMERICA

# ACS Symposium Series

**M. Joan Comstock,** *Series Editor*

# Foreword

The ACS SYMPOSIUM SERIES was founded in 1974 to provide a medium for publishing symposia quickly in book form. The format of the Series parallels that of the continuing ADVANCES IN CHEMISTRY SERIES except that, in order to save time, the papers are not typeset but are reproduced as they are submitted by the authors in camera-ready form. Papers are reviewed under the supervision of the Editors with the assistance of the Series Advisory Board and are selected to maintain the integrity of the symposia; however, verbatim reproductions of previously published papers are not accepted. Both reviews and reports of research are acceptable, because symposia may embrace both types of presentation.

# Contents

# Preface

FOURIER TRANSFORM (FT) TECHNIQUES have been applied to a variety of spectroscopic methods in the past twenty years, as computer technology has rapidly advanced. It was generally believed that mass spectrometry would not benefit from FT techniques because the multiplex or Fellgett's advantage is realized only when the technique is detector-noise limited and not when it is signal-noise limited. In 1974, however, Comisarow and Marshall applied FT techniques to ion cyclotron resonance (ICR) mass spectrometry, which did not employ the same type of detection methods as conventional mass spectrometers. FTICR, or FTMS as it is now commonly called, retains the unique features of conventional ICR, including ion trapping and multiple resonance capabilities, while circumventing its restrictions, including slow scan speeds, low resolution, and limited mass range. As a result, FTMS has evolved into a versatile and powerful spectroscopic technique. In addition, it possesses a number of unique capabilities which give it great potential for becoming a major tool for both analytical applications and basic physical and chemical studies.

The purpose of this symposium was to bring together researchers who are investigating new applications of FTMS, as well as those who are developing instrumental advances in FTMS, in order to report on the current analytical capabilities of FTMS and those projected for the future. The chapters cover a wide variety of applications, ranging from basic studies of photodissociation of ions to the analysis of high molecular weight biopolymers. Although several review articles have appeared in the literature, this is the first book dedicated exclusively to FTMS. It is meant to serve as a general introduction to the technique as well as a summary of current applications. I hope that this book will spark the interest of chemists, biologists, physicists, and others to learn more about FTMS and help them ascertain whether this technique will solve problems in their own laboratories.

I thank the authors who have given their time to participate in the symposium and the publication of this book. Marc Wise, Robert Hettich, and Elizabeth Stemmler of the Organic Spectroscopy Group here at Oak Ridge have also contributed substantially by helping compile the glossary and proof manuscripts; their efforts are greatly appreciated. Finally, I thank Lavonn Golden for secretarial assistance.

MICHELLE V. BUCHANAN
Oak Ridge National Laboratory
Oak Ridge, TN

August 1987

# Chapter 1

# Principles and Features of Fourier Transform Mass Spectrometry

**Michelle V. Buchanan [1] and Melvin B. Comisarow [2]**

[1]Analytical Chemistry Division, Oak Ridge National Laboratory, Oak Ridge, TN 37831-6120
[2]Department of Chemistry, University of British Columbia, Vancouver, British Columbia V6T 1Y6, Canada

Fourier transform mass spectrometry (FTMS) is a rapidly growing technique of increasing analytical importance. Foremost among its many attributes are its high mass resolution and wide mass range capabilities, as well as its ability to store ions. This relatively new technique has been employed in a wide variety of applications, ranging from the exact mass measurement of stable nuclides to the determination of peptide sequences. The future holds considerable promise both for the expanded use of FTMS in a diverse range of chemical problems, as well as advances in the capabilities of the technique itself.

Fourier transform mass spectrometry (FTMS) is an exciting technique that combines the operating features of several different types of conventional mass spectrometers into a single instrument and possesses a number of unique capabilities, as well. Originally developed by Comisarow and Marshall in 1974 (1-3), FTMS is derived from scanning ion cyclotron resonance mass spectrometry (ICR) (4) by the application of Fourier transform (FT) techniques. It should be noted that both in this book and in the general literature, the terms Fourier transform ion cyclotron resonance (FT-ICR) mass spectrometry and Fourier transform mass spectrometry are used synonymously. Although ICR has historically been a valuable tool for the study of gas-phase chemical reactions (5), prior to the introduction of FTMS, the analytical applications of ICR had been restricted by low mass resolution, limited mass range, and slow scanning speeds (6-8). By employing Fourier transform techniques (9) in conjunction with a trapped ion cell (10, 11), FTMS has circumvented these limitations and, in fact, has the potential of becoming an important analytical technique.

This chapter is intended to serve as an overview of the general principles and features of FTMS. Capabilities pertinent to analytical studies will be specifically highlighted. In addition

0097-6156/87/0359-0001$06.00/0

to the papers cited in this chapter, the reader is referred to a number of other reviews (12-16) which provide additional references to detailed reports of applications and research in the field of FTMS.

## Principles of Fourier Transform Mass Spectrometry

Figure 1 is a schematic of a trapped ion cell with cubic geometry, which is commonly used in FTMS (8, 11). The cell is contained within a high-vacuum chamber (pressures of $10^{-6}$ torr or less) which is centered in a homogeneous magnetic field. Magnetic field strengths used for FTMS are typically 1-7 T, with 2-3 T fields generated by superconducting magnets being the most common. Like conventional mass spectrometry, ions may be formed in the FTMS cell by a number of methods, including electron impact (1), chemical ionization (17-20), laser ionization and desorption (21-23), and particle induced desorption, such as secondary ion mass spectrometry and plasma desorption (24-27). After formation, ions are trapped in the cell, held in the radial direction (xy plane) by the magnetic field and along the axis of the magnetic field (z-axis) by small voltages (0.5 to 5 V) applied to the trapping plates. Either positive or negative ions may be trapped in the cell simply by changing the polarity of the voltage applied to the cell plates.

The frequency of the cyclic motion of ions, $\omega$, within the cell is given by the cyclotron equation:

$$\omega = KqB/m \qquad (1)$$

where K is a proportionality constant, q is the charge of the ion, m is its mass, and B is the magnetic field strength. Because the magnetic field strength is constant in the FTMS experiment, ions of different mass will have unique cyclotron frequencies. For example, at a magnetic field strength of 3 T, an ion with a mass to charge ratio (m/z) of 18 will have a cyclotron frequency of 2.6 MHz, while an ion at m/z 3,000 will have a frequency of 15.6 KHz.

Because of momentum conservation, the initial ion velocity upon ion formation is the same as the velocity of its neutral precursor. For a macroscopic ensemble of ions, there is no net coherent cyclotron motion even though the ion orbits are nonzero. Without coherent motion, a signal cannot be detected. By applying a very short, high intensity, broadband radiofrequency signal (2) ("chirp") to the excite (or transmitter) plates of the cell, the ions absorb energy, which accelerates them into larger orbits and causes them to move together (coherent motion). The orbiting packet of ions induces a small alternating current ("image current") in the receiver plates (28). This signal is converted into a voltage, amplified, digitized and stored in a computer (28). The frequency components of the image current correspond to the cyclotron frequencies of the ions present in the cell. If ions of only one mass to charge ratio were present in the cell, the detected signal (time domain signal) would resemble a single frequency sine wave. However, in the FTMS experiment, all ions in the cell are excited virtually simultaneously (3) and detected simultaneously (2, 3, 7). The resulting time domain spectrum is very complex because the signals of all ions are superimposed. In order to recover the frequency information, the complex time-domain spectrum is subjected to a Fourier transform algorithm. This generates a frequency domain spectrum, which can be readily converted to the familiar mass spectrum using the cyclotron relationship given above, Equation 1 (3, 29, 30). Another advantage of the Fourier transform technique is that data acquisition time is greatly reduced because ions of all masses are detected simultaneously (7).

Because frequency can be so precisely measured, the exact mass of an ion can be determined very accurately in the FTMS experiment (8). Typically, low parts-per-million accuracy can be achieved in the presence or even in the absence of an internal mass calibrant (13). In addition, a high degree of mass accuracy can be maintained for days without recalibration provided that the magnetic field remains stable. More detailed information on the theory of FTMS (1, 16, 28, 31-33) and the principles of Fourier transforms applied to spectroscopic techniques (9, 34) may be found in the literature.

### The Basic FTMS Experiment

In practice, the basic FTMS experiment is conducted using a series of computer-controlled pulse sequences, as depicted in Figure 2. In this experiment, ion formation and detection are separated in time. This is in contrast to conventional mass spectrometers, which rely on spatial dispersion of ions. In the simplest case, the initial pulse involves the formation of ions, whether by electron ionization or another technique. The second is the frequency sweep (2) excitation pulse which brings the ions into coherent motion, as described previously. This is followed by a detection period during which the signal (image current) (28) from the ions is received. A quench pulse is then applied to the cell to remove all the ions from the cell prior to starting the next pulse sequence. The entire experiment sequence can take less than a second (a minimum of about 10 ms for an electron ionization experiment), and the series of pulses may be repeated as many times as desired for signal averaging and enhanced signal-to-noise ratio. Additional pulses or delays between pulses may be inserted in this basic pulse sequences in order to perform more complex experiments, as will be discussed later.

In addition to the cubic cell geometry (11) depicted in Figure 1, a variety of other cell geometries have been devised (4, 35, 36). One that has attacted considerable interest for analytical purposes is the dual cell (37), depicted in Figure 3. In this design, two differentially pumped cubic cells share a common trapping plate which contains a small (2 mm) orifice. This arrangement allows the cells to be operated at a pressure differential of $10^3$. Ions may therefore be formed at higher pressures in the "source" cell and transferred to the "analyzer" cell for detection under low pressure conditions. In the FTMS experiment, low pressures are required for high resolution mass measurement and other high performance capabilities, as outlined below.

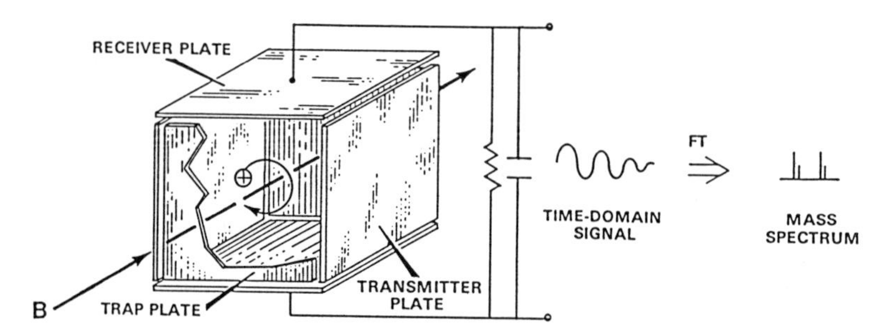


Figure 1. Schematic representation of a cubic trapped ion cell commonly used in FTMS. Coherent motion of ions in the cell induces an image current in the receiver plates. The time domain signal is subjected to a Fourier transform algorithm to yield a mass spectrum.

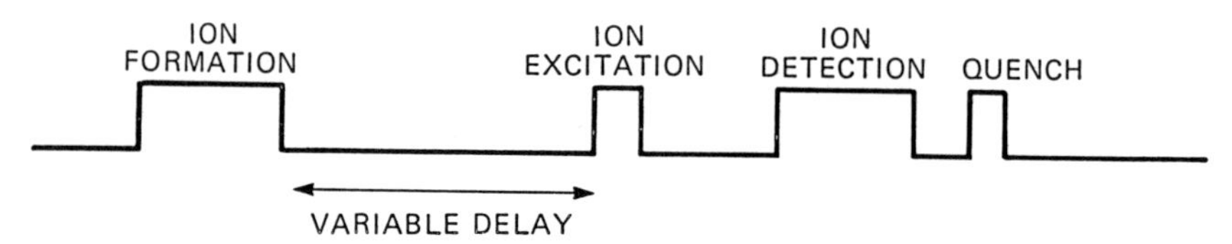


Figure 2. Simple pulse sequence used in elementary FTMS experiments, including ion formation, excitation, and detection. A quench pulse is used to eject ions from the cell prior to repeating the sequence. The variable delay between ion formation and excitation can be used for ion storage and manipulation, as described in the text.

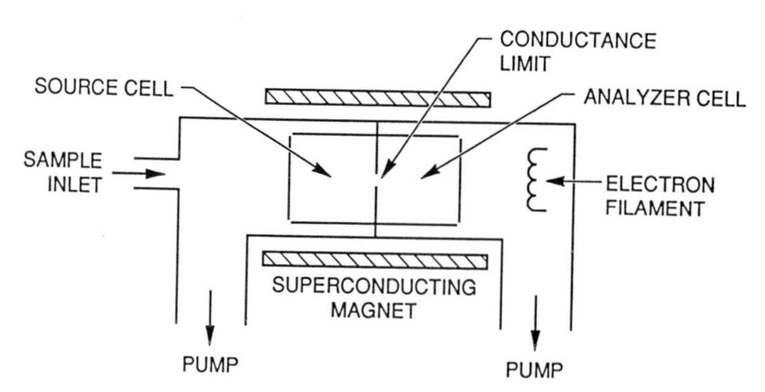


Figure 3. Schematic representation of a dual cell used for FTMS. The two differentially pumped cells are joined by a trapping plate containing a small orifice that is capable of supporting a $10^3$ pressure differential. Ions may be detected in either cell. However, if desired, ions formed in the "source" region under higher pressure conditions may be transferred to the "analyzer" cell for detection under lower pressure conditions, as described in the text.

Features of FTMS

FTMS is mechanically simple, yet remarkably versatile. The instrument can be operated in any number of modes simply by changing the pulse sequences used to control the spectrometer. As a result, a wide variety of experiments can be performed with a sample in a very short period of time. For example, by altering just a few parameters using the computer keyboard, one may switch from positive to negative ion detection, from electron ionization to chemical ionization, and from a single stage experiment to a multiple stage experiment (e.g., MS/MS). In addition, a number of features makes FTMS particularly well-suited for analytical studies, including high mass resolution and wide mass range, ability to store, manipulate and selectively detect ions, and compatibility with a variety of ionization techniques. Examples of these features are given below and additional examples may be found in previous reviews of analytical applications of FTMS (13-15, 38).

High resolution and wide mass range. One of the most outstanding features of FTMS is its ability to achieve very high mass resolution, which was predicted early in the development of the technique (7, 39). Mass resolution in the FTMS varies inversely with mass and increases proportionally to observation time and magnetic field strength (8, 40). An increase in operating pressure causes collisional damping of the coherent motion of the ions within the cell and a rapid decay of the image current signal, resulting in a decrease in resolution. Therefore, very low pressure ($10^{-8}$ torr or lower) is required in order to monitor the signal for a sufficiently long period of time to obtain high mass resolution.

Mass resolution in excess of 500,000 (FWHM) at mass 100 can be readily obtained with most commercial FTMS instruments and resolution of 200,000,000 at m/z 40 (41) has been reported. This compares with the highest mass resolution of commercial sector instruments of about 120,000. In addition, high resolution measurements can be made in a matter of seconds with an FTMS and the instrument can be switched between medium and high resolution modes simply by changing a few parameters through the computer. With conventional magnetic sector instruments, which must be mechanically adjusted (14), this change can be very time-consuming. Further, with conventional mass spectrometers, high mass resolution is achieved by narrowing the ion transmission slits, which results in a decrease in the signal. In FTMS, increased resolution is achieved simply by accumulating the signal for longer periods of time and results in an increase in the signal to noise ratio (S/N) as well (42). Thus, high resolution spectra can often be obtained on very small quantities of sample using FTMS.

In addition to high mass resolution, another important feature of FTMS is its wide mass range (7,8). From examination of the cyclotron equation (Eq. 1), the mass range of FTMS appears to have no upper limit. However, an instrumental upper limit in excess of 100,000 has been suggested (16), based on a detailed study of ion motion (30). From a practical standpoint, the cyclotron frequencies of high mass ions (> 5,000 daltons) are low enough that environmental (1/f) noise can severely interfere with their detection. Improved electronics have been devised to help circumvent this problem, and ions at mass 16,241 from cesium iodide clusters have been detected using an FTMS equipped with a 3 T magnet (43). Alternatively, as higher field magnets are developed and employed, the upper mass range of FTMS will be extended as well. The low mass limit is governed by the rate of signal digitization. For example, according to Nyquist theory, if a 5.2 MHz digitizer is used, the highest frequency signal which can be processed without distortion is 2.6 MHz. As previously mentioned, at a field strength of 3 T, this frequency corresponds to about 18 daltons. Lower masses can be obtained, however, by use of a faster signal digitizer or with lower magnetic fields. Alternatively, a special narrow-band excitation (heterodyne) technique which reduces the digitization requirements of the signal can be used to obtain lower masses (7,8).

The combination of high mass resolution, wide mass range, and exact mass capabilities makes FTMS an important tool for determining the chemical formula of an unknown molecule. This can be especially important for higher molecular weight compounds such as polymers and biological molecules. Resolution of 150,000 at m/z 1180 has been reported (37). High resolution is also required for the separation of isobaric ions and can be of importance in the analysis of mixtures. For example, Figure 4 is a spectrum obtained from cigarette smoke tar, which contains many thousands of compounds. The sample was deposited on a direct insertion probe, and the spectrum was collected using medium resolution conditions. Closer inspection of the area around mass 124 (expansion shown in Figure 5) reveals three well-resolved isobaric ions. Under the moderate resolution conditions used to acquire the spectrum (resolution of 24,000 at m/z 124) molecular formulae for the three compounds can readily be established. Using the high resolution capabilities of FTMS, the same approach can be used to identify isobaric species at much higher masses.

The requirement of low pressure for high resolution mass measurement can limit the performance of FTMS when used in conjunction with chromatographic techniques, such as gas chromatography or supercritical fluid chromatography, or with ionization techniques that require higher source pressures, such as fast atom bombardment or high pressure chemical ionization. Using a jet separator to reduce the pressure within the ion cell, mass resolution for benzene (m/z 78) of 24,000 has been obtained, compared to 8000 without the jet separator (44). Pulsed valves have also been used to limit the gas load in the ion cell by admitting the GC effluent only during the ionization event and allow the gas to be pumped away prior to the ion detection step (45).

Separation of the ionization region from the analyzer region is another means of increasing mass resolution by reducing the pressure in the analyzer region and has the advantage of retaining high mass resolution with little or no loss sensitivity (12). This separation can be accomplished using a tandem quadrupole-FTMS arrangement (46, 47), an external ionization cell (48, 49) or the

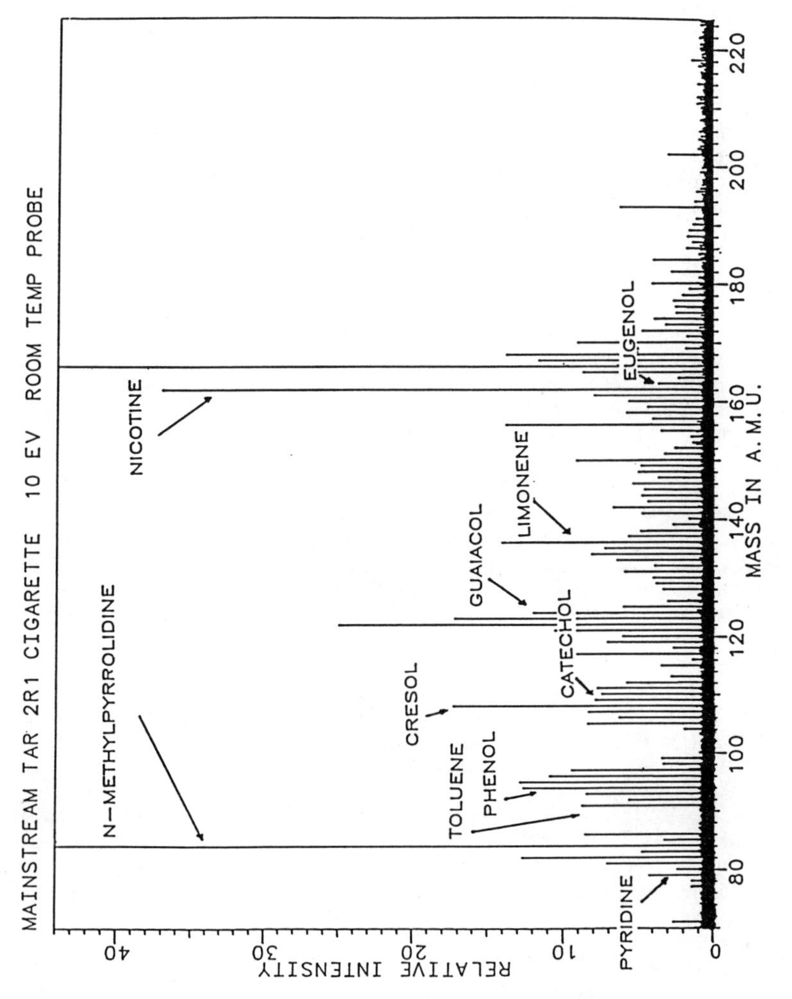


Figure 4. Medium resolution low energy (10 eV) electron ionization spectrum of tobacco smoke tar desorbed from a heated solids probe.

dual cell design (37) described earlier. Using a differentially pumped dual cell interfaced with a GC, mass resolution of 30,000 at m/z 174 has been reported (37). This is sufficient to allow chemical formulae of GC peaks to be determined. Recently, supercritical fluid chromatography has been successfully interfaced to an FTMS equipped with a dual cell (50).

Ion storage and manipulation. Once formed, ions may be trapped within the ion cell for as long as 5 x $10^4$ seconds (51) prior to detection. During this time and prior to the excitation pulse and subsequent detection, a number of processes can be used to probe molecular structure. For example, if a short delay (tens to hundreds of msec) is introduced after ion formation, the low mass fragment ions formed from a molecule can react with neutrals to yield ionic products (17, 52). These ions, in turn, can serve as proton transfer agents to ionize remaining neutrals (18). This phenomenon has been called "self-CI" or reagentless CI because no reagent gas is required as in conventional chemical ionization (CI) techniques.

Low pressures ($10^{-6}$ torr) of a reagent gas may be introduced during the delay between ion formation and excitation to provide more selective CI reactions, such as hydrogen/deuterium exchange reactions (18) or to provide a source of low energy electrons for electron capture negative ion CI reactions (53, 54). As in the case of "self-CI", the extent of these CI reactions may be controlled by changing the delay time between ion formation and excitation.

CI reactions using reagent gas pressures of greater than $10^{-6}$ torr have also been performed using pulsed valve introduction of the reagent (55), in a manner similar to pulsed valve GC/FTMS discussed previously. The CI products can be detected with higher mass resolution because the reagent gas is pumped away prior to the detection step. The dual cell (37) and external source designs (35, 46-48) are particularly well-suited for high pressure CI reactions. In the case of the dual cell, the reagent gas may be introduced into the source cell using either pulsed valves or a static pressure of reagent gas. After a specified delay time to allow the CI reactions to occur, the product ions are transferred to the lower pressure analyzer cell, where they are detected.

Chemical ionization reactions using so-called non-conventional reagents have also been performed using FTMS. These would include compounds with limited volatility that are not amenable for use in conventional high pressure CI sources. Chemical ionization in the FTMS experiment is accomplished by multiple collisions obtained by introducing longer delay times after ion formation, eliminating the need for high concentration of CI reagent ions. For example, metal ions can be readily formed by laser impact on metal targets (56). These ions have been shown to react selectively with organic compounds and have great potential for providing structural information (57, 58).

Ions can also be manipulated after formation by selective excitation using a radiofrequency pulse. A high-power rf pulse may be used to excite an ion (or group of ions) to a sufficiently large

orbital(s) to cause neutralization on the cell plates (52).
process, called double resonance (52, 59), has a numbe
applications. For example, suspected reactant ions may
selectively ejected from the cell using double resonance to confi
their participation in observed ion-molecule reactions (17, 52)
Another important application of double resonance is to increase
the dynamic range of FTMS, which is generally limited to about $10^3$ to $10^4$. By ejecting a major ion in a spectrum, the dynamic range of the measurement can be increased, allowing the less abundant components to be observed.

Double resonance can also be used to isolate ions of a single mass in the cell so that additional gas phase reactions may be performed. An important example of this type of process is collisional activation (60, 61), or MS/MS, a technique of increasing analytical importance (62, 63). After the ion of interest is isolated, a second radiofrequency pulse (lower in energy than the ion ejection pulse) is used to excite the ion to a higher kinetic energy. The excited ion undergoes collisions with a suitable target gas, such as argon, which causes the ion to fragment. The mass spectrum of these fragment ions is characteristic of the original ion. At present, the resolution with which a parent ion may be selected is generally limited to unit mass. However, as will be discussed later, new excitation techniques should overcome this limitation.

With conventional MS/MS instruments, several types of mass analyzers are linked together to perform the isolation and analysis steps (62). FTMS performs these steps in one trapped ion cell. In addition, the isolation and excitation steps may be repeated to obtain $MS^n$. Recently, four stages of MS/MS or $MS^5$ has been reported using FTMS (64). The ultimate limitation of the number of MS/MS stages possible by FTMS is loss of the ion signal (due to collisional inefficiencies, ion-molecule reactions, and ion ejection), not hardware or software (15). Therefore, in extended multiple MS/MS experiments, such as the $MS^5$ experiment, it is advantageous to utilize conditions that minimize the number of different daughter ions produced. Extensive fragmentation reduces the percentage of the ion current carried by the daughter ion to be collisionally dissociated in the next step.

FTMS has several advantages for selected MS/MS experiments. First, the energy of the collision can be varied with the power of the excitation and is proportional to $B^2r^2$, where B is the magnetic field strength and r is the radius of the cyclotron motion. The ability to alter the energy of the collision can be of particular use in structural studies because collisional spectra are sensitive to energy (60). Typical collisional energies used in FTMS range up to a few hundred electron volts, which is considerably lower than for conventional tandem sector instruments, which can attain several thousand electron volts. In addition, the collisional energies obtainable with FTMS decrease as mass increases (15). However, by using larger cells and greater magnetic field strength, higher collisional energies can be acquired. For example, by employing a larger cell, high-energy charge stripping of benzene has been performed (65).

of FTMS for collisional dissociation is that d by multiple collisions, which can enhance rangement processes. After excitation, the a finite kinetic energy distribution and kinetic energy is greater at lower excitation because of this, the daughter ion spectra obtained n be less reproducible than with higher energy sector ics.

Another advantage of FTMS for MS/MS experiments is that high resolution daughter ion spectra can be obtained. Using a single cell and narrow band (heterodyne)(7, 17) detection, daughter ion resolutions of several thousand have been reported (67, 68). A ten-fold increase in resolution for daughter ion spectra was reported when pulsed valves were used to introduce the collision gas and allow it to be pumped away prior to mass analysis (12). Use of a differentially pumped dual cell has allowed much higher resolution daughter ion spectra to be obtained (69). Figure 6 shows a spectrum obtained on the m/z 105 ions obtained from the dissociation of the m/z 120 ions from 1,3,5-trimethylbenzene (mesitylene) and acetophenone. The parent ions were formed by electron impact at 10 eV in the high pressure source cell, and ions of lower mass were ejected by double resonance pulses. The m/z 120 ions were excited to an energy of 125 eV (laboratory frame of reference) and collided with argon, present in the source cell at a pressure of $6 \times 10^{-6}$ torr. The daughter ions were then pulsed across the conductance limit into the lower pressure ($1 \times 10^{-8}$ torr) analyzer cell, where they were detected with a resolution of 211,000 (FWHM). Similar experiments with negative ions have yielded daughter ion spectra with resolution in excess of 320,000 (FWHM) at m/z 46 from 2,4-dinitrotoluene.

Other means of manipulating ions trapped in the FTMS cell include photodissociation (70-74), surface induced dissociation (75) and electron impact excitation ("EIEIO")(76) reactions. These processes can also be used to obtain structural information, such as isomeric differentiation. In some cases, the information obtained from these processes gives insight into structure beyond that obtained from collision induced dissociation reactions (74). These and other processes can be used in conjunction with FTMS to study gas phase properties of ions, such as gas phase acidities and basicities, electron affinities, bond energies, reactivities, and spectroscopic parameters. Recent reviews (4, 77) have covered many examples of the application of FTMS and ICR, in general, to these types of processes. These processes can also be used to obtain structural information, such as isomeric differentiation.

<u>Ionization of Non-volatile Samples.</u> In addition to electron impact ionization and chemical ionization, which were discussed earlier, FTMS can be used in conjunction with a variety of other ionization techniques. The wide mass range and high mass resolution capabilites of FTMS make it particularly well-suited for the analysis of high molecular weight compounds, such as polymers and biological samples. These compounds typically have limited volatility due to high molecular weight and/or polarity and are

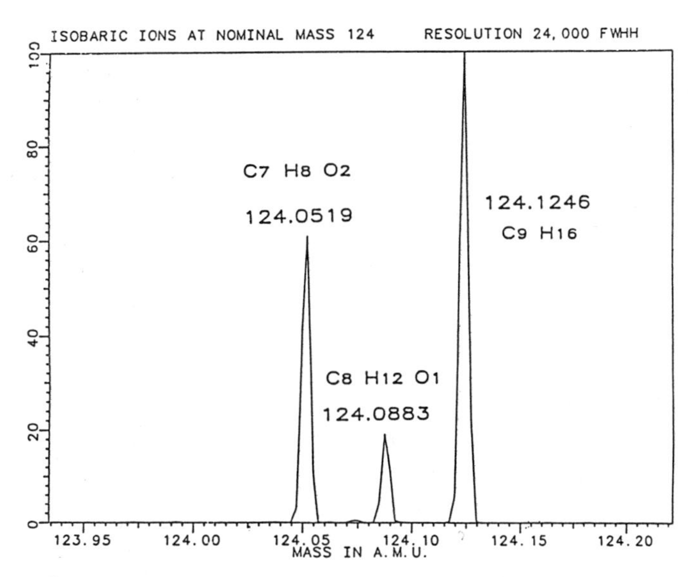


Figure 5. Expansion of the region around m/z 123 of Figure 4 showing three well-resolved isobaric ions.

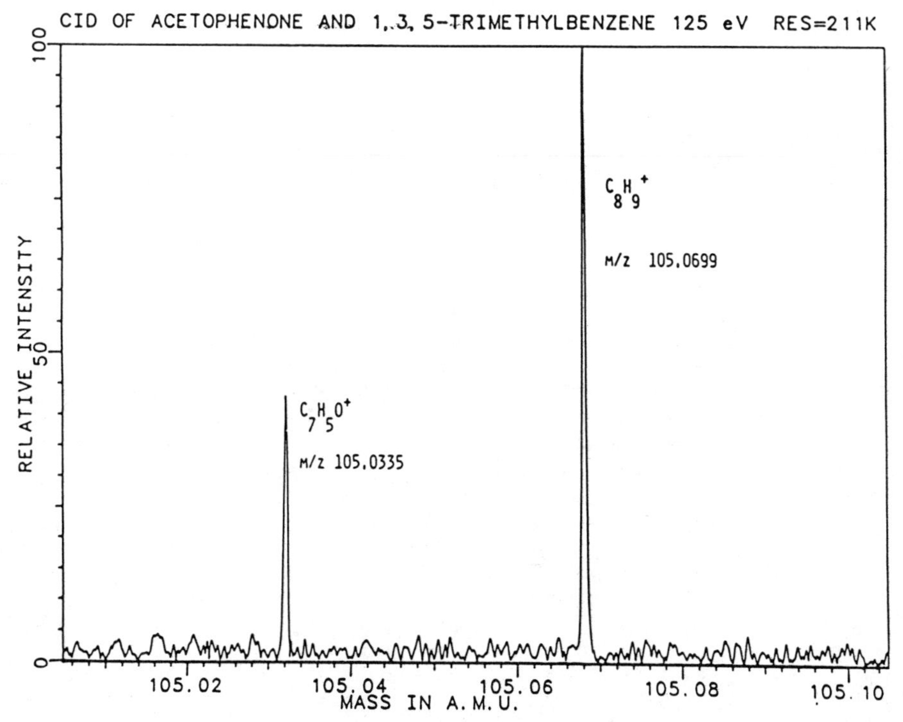


Figure 6. High-resolution CAD daughter ion spectrum of a mixture of acetophenone and 1,3,5-trimethylbenzene from m/z 120 parent ions.

difficult to ionize using conventional techniques. A number of so-called "soft ionization" techniques have been developed for the analysis of high molecular weight compounds by mass spectrometry, including laser desorption, secondary ion methods, and plasma desorption. All of these techniques are applicable to FTMS. In addition, field desorption, one of the first "soft ionization" techniques, has recently been used with FTMS (78).

The combination of FTMS and lasers (79) is particularly well-suited for ionization of non-volatile samples (80) because the FTMS experiment is amenable to non-continuous ion production provided by pulsed lasers. Further, laser desorption offers a means of generating high mass ions with a minimum of fragmentation. After ionization, the versatile features of FTMS can be used to obtain structural information on these high mass ions.

Metal ions have been produced from metal targets using a pulsed Nd/YAG laser and reacted with organics for subsequent study by FTMS (21, 81, 82). Further, cluster ions can be formed by laser desorption and subsequently studied by FTMS (49, 83). In an early study, FTMS spectra of polar and zwitterionic compounds were obtained using a pulsed TEA $CO_2$ laser (22). Since that time, pulsed lasers have been employed for the ionization of non-volatiles in a number of studies (19, 37, 84-86). For example, the molecular weight distributions of polymers with average molecular weights of up to 6000 daltons have been determined using a pulsed $CO_2$ laser (87). Solutions of the polymers were doped with either KBr or NaCl and the resulting spectra revealed no significant fragmentation, simplifying molecular weight characterization.

Laser desorption has been employed to ionize nucleosides, oligosaccharides and glycosides, generating both positive and negative ions in the resulting FTMS spectra (88). Laser desorption FTMS has also been used to obtain sequence information on mixtures of peptides (89). Selective excitation of the molecular ions produced by laser desorption was followed by collisional activation to yield complete sequence information on a cyclic decapeptide and for 12 of 15 amino acids of a linear peptide. Oligonucleotides have been studied using laser desorption FTMS and collisional activation, as well (90). Figure 7 shows a negative ion daughter spectrum obtained from the adenine base originally contained in the tetranucleotide d(AGCT), or deoxy(adenosine-guanosine-cytidine-thymidine). This tetranucleotide, which has a molecular weight of 1173 daltons, was ionized using the fundamental line of a Nd/YAG laser at 1064 nm. The fragment anion at m/z 134 was isolated and identified as adenine, based on collisional dissociation, forming a daughter ion at m/z 107, which corresponds to a loss of HCN.

Structural information on ions produced by laser desorption has also been obtained by using infrared multiphoton dissociation processes (73). A pulsed $CO_2$ laser was used to form the ions, which were further fragmented either by using sequential pulses from the same laser or by using a gated continuous wave infrared laser.

Laser desorption also produces neutral species which can be subsequently ionized by electron ionization or by "self-CI" processes. Laser desorption/electron ionization has been used to

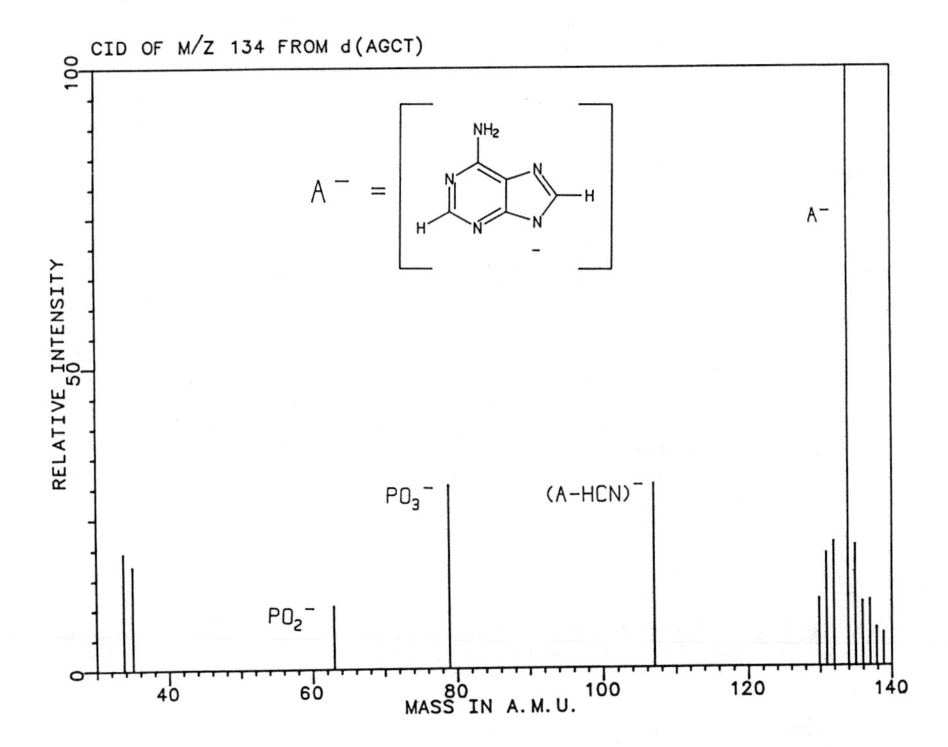


Figure 7. Daughter ion spectrum from the m/z 134 anion isolated from the tetranucleotide deoxy(adenosine-guanosine-cytidine-thymidine). Ions were generated by laser desorption from a direct insertion probe.

ionize erythromycin (91) and produced a very clean spectrum with about two orders of magnitude less sample than fast atom bombardment (FAB). In addition, greater molecular weight information was obtained with the LD/EI/FTMS spectrum than with the FAB spectrum.

Particle induced desorption methods are commonly used to ionize low-volatility compounds. Cesium ion desorption (or cesium ion secondary ion mass spectrometry, SIMS) uses a primary beam of cesium ions to desorb and ionize a non-volatile sample. This technique has been used with FTMS to produce pseudomolecular ions of vitamin $B_{12}$, $\{(B_{12})_2 + Cs - 2CN\}^+$, at m/z 2792 (24) and molecular ions of **beta**-cyclodextrin (m/z 1135) (92, 93). Detection limits of $10^{-13}$ mol for the peptide gramicidin S has been demonstrated using $Cs^+$ SIMS with FTMS (25), and additional structural information was obtained using MS/MS processes.

Fast atom bombardment has been used with an FTMS instrument equipped with an external source (46, 94), and protonated molecules of cytochrome C at m/z 12,385 have been observed. Recently, $^{252}Cf$ plasma desorption (PD) (95) has been used in conjunction with FTMS (27, 75, 96, 97), and spectra of compounds with molecular weights up to 2000 daltons have been observed (27, 75). This technique has generally been used only with time-of-flight (TOF) mass spectrometers because the ion currents obtainable are typically too low for scanning instruments. PD ionization with FTMS offers a number of advantages over PD/TOF, including higher mass resolution and the capability of ion manipulation to probe molecular structure.

## Future Developments and Applications of FTMS

Two important features of FTMS that will be widely exploited in the future are the ultra-high mass resolution and the wide mass range of the technique. It should be noted that to a considerable extent, the need for greater mass range and higher mass resolution has been stimulated by the development of specialized ionization techniques such as laser desorption, fast atom bombardment, plasma desorption, and secondary ion (SIMS) methods. As these and other processes for the production of gas phase ions from involatile solid samples are developed and refined, FTMS will be exceptionally useful for the mass analysis of these materials. For example, the analysis of biopolymers by mass spectrometry is an area that is growing rapidly at present (98). For ordinary protein and nucleotide sequencing, mass spectrometry can be used as an adjunct to traditional biochemical techniques. However, for modified or N-terminus blocked proteins, for which the conventional Edman degradation method often fails, mass spectrometry will be important in its own right (98). Similarly, FTMS could be used to sequence modified nucleotides.

The ultra-high mass resolution capabilities of FTMS will also be used in the future for accurate determination of nuclide masses. It is possible that all the known stable nuclides will have their masses reexamined by FTMS. A recent example of this type of measurement is the determination of the difference in mass between $^3He^+$ and $^3H^+$ by FTMS (99), which may be used in conjunction with other data to establish the mass of the antineutrino. Another interesting use of ultra-high resolution FTMS is the direct determination of the relativistic mass difference between the ground state and metastable excited state of ions, such as $Ar^+$ (100).

FTMS also has the potential of becoming an important tool for determining molecular structure. Traditionally, mass spectrometry has been rather limited in its ability to determine the structure of an unknown compound unambiguously. Additional structural methods, such as nuclear magnetic resonance or crystallography, are commonly used in conjunction with mass spectrometry to elucidate the identity of a molecule. However, when the amount of sample is severely limited or when the sample is a component in a complex mixture, mass spectrometry is often one of the few analytical techniques that can be used.

The ability to trap and manipulate ions in the FTMS makes this a potentially powerful tool for structural determination. The FTMS has been described as a "complete chemical laboratory" (101, 102), where reactions can be used to "pick apart" a molecule systematically using sequential CAD, photodissociation, chemical reactions, or other techniques. As selective and sensitive processes for these reactions are developed, FTMS has the potential of yielding detailed information on the structure of a molecule which is currently only obtainable using techniques that require considerably larger sample sizes. It should also be noted that reactions of trapped ions with neutrals can be also be devised for the step-wise synthesis of a particular species in the FTMS (102, 103).

As noted earlier, the development of the dual cell (37), tandem quadrupole-FTMS (46, 47) and external ionization cell (48, 49) has facilitated the coupling of FTMS and chromatographic methods. Advances in interfacing separation techniques with FTMS will be important in the analysis of mixtures, especially where high mass resolution is required. For example, liquid chromatographic introduction of mixtures isolated from biological systems directly into an FTMS for analysis would eliminate the need for laborious sample clean up.

The future of FTMS from an instrumental standpoint also shows considerable potential. The performance of any FTMS instrument is dependent upon the performance of the computer, magnet, and the vacuum system. While no substantial improvements are on the horizon for the price/performance ratio of vacuum components, the price/performance ratio of computers and magnets improve each year. It is reasonable to expect that the steadily decreasing price of computers and magnets will lower the cost of FTMS instruments and promote their more widespread application. In addition, with the recent surge of research in higher temperature superconducting magnets, it is possible that much smaller, less expensive, and easier to maintain magnets might be available in the future. The commercial development of small, lower performance FTMS instruments based on lower field magnets or other means of ion trapping are also a possibility in the future.

Finally, in addition to computers becoming more powerful, faster, and less expensive, new mathematical- and computer-based concepts are being developed to increase the capabilities of FTMS. One example, which is outlined in a subsequent chapter in this volume (104), is Stored Waveform Inverse Fourier Transform techniques, or SWIFT. These new techniques have the potential to reduce substantially the previous limitations of FTMS with respect to areas such as dynamic range, accurate isotopic ratios, and high resolution parent ion selection in MS/MS experiments.

The high performance capabilities of FTMS, combined with the ability to manipulate trapped ions, makes the FTMS an extremely versatile and powerful tool. The future promises to bring even greater capabilities to this relatively new technique. More applications of FTMS to diverse fields will be seen, as well as the maturation of FTMS into a routine analytical technique.

Acknowledgments

Research sponsored (MVB) jointly by the National Cancer Institute under Interagency Agreement No. 0485-0485-A1 and the Office of Health and Environmental Research, U.S. Department of Energy under contract DE-AC05-84OR21400 with Martin Marietta Energy Systems, Inc. Support is also gratefully acknowledged (MBC) from the Natural Sciences and Engineering Research Council of Canada.

Literature Cited

1. M. B. Comisarow and A. G. Marshall, Chem. Phys. Lett. (1974) **4**, 282-293.
2. M. B. Comisarow and A. G. Marshall, Chem. Phys. Lett. (1974) **26**, 489-490.
3. M. B. Comisarow and A. G. Marshall, Can. J. Chem. (1974) **52**, 1997-1999.
4. K. P. Wanczek, Int. J. Mass Spectrom. Ion Proc. (1984) **60**, 11-60.
5. J. E. Bartmess and R. T. McIver, Jr., in "Gas-Phase Ion Chemistry", ed. M. T. Bowers, Academic Press, New York, 1979, Vol. 2, p. 81-121.
6. J. M. H. Henis, Anal. Chem. (1969) **41**, 22-32A.
7. M. B. Comisarow, Adv. Mass Spec. (1978) **7**, 1042-1046.
8. M. B. Comisarow, Adv. Mass Spectrom. (1980) **9**, 1698.
9. A. G. Marshall, in "Fourier, Hadamard, and Hilbert Transforms in Chemistry", ed. A. G. Marshall, Plenum Press, New York, 1982, pp 1-43.
10. R. T. McIver, Jr., Rev. Sci. Instrum. (1970) **41**, 555.
11. M. B. Comisarow, Int. J. Mass Spectrom. Ion Phys. (1982) **37**, 251-257.
12. A. G. Marshall, Acc. Chem. Res. (1985) **18**, 316.
13. C. L. Johlman, R. L. White, and C. L. Wilkins, Mass Spectrom. Rev. (1983) **2**, 389.
14. C. L. Wilkins and M. L. Gross, Anal. Chem. (1981) **53**, 1661A.
15. M. L. Gross and D. L. Rempel, Science (1984) **226**, 261.
16. M. B. Comisarow, Anal. Chim. Acta (1985) **178**, 1.
17. G. Parisod and M. B. Comisarow, Adv. Mass Spectrom. (1980) **8A**, 212-223.

18. S. Ghaderi, P. S. Kulkarni, E. B. Ledford, and M. L. Gross, Anal. Chem. (1981) **53**, 428.
19. T. J. Carlin and B. S. Freiser, Anal. Chem. (1983) **55**, 571.
20. D. A. McCrery, T. M. Sack, and M. L. Gross, Spectrosc. Int. J. (1984) **3**, 57.
21. R. B. Cody, R. C. Burnier, and B. S. Freiser, Anal. Chem. (1982) **54**, 96.
22. D. A. McCrery, E. B. Ledford, Jr., and M. L. Gross, Anal. Chem. (1982) **54**, 1435.
23. M. P. Irion, W. D. Bowers, R. L. Hunter, R. S. Rowland, and R. T. McIver, Jr., Chem. Phys. Lett. (1982) **93**, 375.
24. M. E. Castro and D. H. Russell, Anal. Chem. (1984) **56**, 578.
25. I. J. Amster, J. A. Loo, J. J. P. Furlong, and F. W. McLafferty, Anal. Chem. (1987) **59**, 313.
26. D. F. Hunt, J. Shabanowitz, R. T. McIver, R. L. Hunter, and J. E. P. Syka, Anal. Chem. (1985) **57**, 765.
27. J. A. Loo, E. R. Williams, I. J. Amster, J. J. P Furlong, B. H. Wang, F. W. McLafferty, B. T. Chait, and F. H. Field, Anal Chem. (1985) **57**, 1880.
28. M. B. Comisarow, J. Chem. Phys. (1978) **69**, 4097.
29. E. B. Ledford, Jr., S. Ghaderi, R. L. White, R. B. Spenser, P. S. Kulkarni, C. L. Wilkins, and M. L. Gross, Anal. Chem. (1980) **52**, 463.
30. E. B. Ledford, D. L. Rempel, and M. L. Gross, Anal. Chem. (1984) **56**, 2744.
31. A. G. Marshall, M. B. Comisarow, and G. Parisod, J. Chem. Phys. (1971) **71**, 4434.
32. A. G. Marshall, Chem. Phys. Lett. (1979) **63**, 515.
33. A. G. Marshall and D. C. Roe, J. Chem. Phys. (1980) **73**, 1581.
34. P. R. Griffiths, "Fourier Transform Techniques in Chemistry", Plenum Press, New York, 1978.
35. R. P. Grese, D. L. Rempel, and M. L. Gross, this volume.
36. F. H. Laukein, M. Allemann, P. Bischofberger, P. Grossmann, Hp. Kellerhals, and P. Kofel, this volume.
37. R. B. Cody, J. A. Kinsinger, S. Ghaderi, I. J. Amster, F. W. McLafferty, and C. E. Brown, Anal. Chim. Acta (1985) **178**, 43.
38. E. B. Ledford, Jr., S. Ghaderi, C. L. Wilkins, and M. L. Gross, Adv. Mass Spectrom. (1981) **8B**, 1707.
39. M. B. Comisarow and A. G. Marshall, J. Chem. Phys. (1975) **62**, 293.
40. M. B. Comisarow and A. G. Marshall, J. Chem. Phys. (1976) **64**, 110.
41. K. -P. Wanczek, 1987 Pittsburgh Conference on Analytical Chemistry and Applied Spectroscopy, Atlantic City, NJ, paper No. 526.
42. R. L. White, E. B. Ledford, Jr., S. Ghaderi, C. L. Wilkins, and M. L. Gross, Anal. Chem. (1980) **52**, 1525.
43. I. J. Amster, F. W. McLafferty, M. E. Castro, D. H. Russell, R. B. Cody, Jr., and S. Ghaderi, Anal. Chem. (1986) **58**, 458.
44. E. B. Ledford, Jr., R. L. White, S. Ghaderi, C. L. Wilkins, and M. L. Gross, Anal. Chem. (1980) **52**, 2450.

45. T. M. Sack and M. L. Gross, Anal. Chem. (1983) **55**, 2419.
46. D. F. Hunt, J. Shabanowitz, R. T. McIver, Jr, R. L. Hunter, and J. E. P. Syka, Anal. Chem. (1985) **57**, 765.
47. R. T. McIver, Jr., R. L. Hunter, and W. D. Bowers, Int. J. Mass Spectrom. Ion Proc. (1985) **64**, 67.
48. P. Kofel, M. Allermann, Hp. Kellerhals, and K. P. Wanczek, Int. J. Mass Spectrom. Ion Proc. (1985) **65**, 97.
49. J. M. Alford, P. E. Williams, and R. E. Smalley, Int. J. Mass Spectrom. Ion Proc. (1986) **72**, 33-51.
50. R. B. Cody, Jr. and J. A. Kinsinger, this volume.
51. M. Allemann, Hp. Kellerhals, and K. P. Wanczek, Chem. Phys. Lett. (1980) **75**, 328.
52. M. B. Comisarow, V. Grassi, and G. Parisod, Chem. Phys. Lett. (1978) **57**, 413.
53. M. V. Buchanan, I. B. Rubin, and M. B. Wise, Biomed. Environ. Mass Spectrom. (1987) **14**, 395.
54. M. V. Buchanan and M. B. Wise, this volume.
55. T. M. Sack and M. L. Gross, abstract of paper presented at 31st Annual Conference of Mass Spectrometry and Allied Topics, Boston, May 1983, p. 396.
56. R. A. Forbes, E. C. Tews, B. S. Freiser, M. B. Wise, and S. P. Perone, Anal. Chem. (1986) **58**, 684.
57. R. C. Burnier, G. D. Byrd, and B. S. Freiser, Anal. Chem. (1980) **52**, 1641.
58. D. A. Peake and M. L. Gross, Anal. Chem. (1985) **57**, 115.
59. L. R. Anders, J. L. Beauchamp, R. C. Dunbar, and J. D. Baldeschwieler, J. Chem. Phys. (1966) **45**, 1062.
60. F. W. McLafferty, Science (1981) **214**, 1981.
61. R. T. McIver, Jr. and W. D. Bowers, in "Tandem Mass Spectrometry", ed. F. W. McLafferty, Wiley, New York, 1983, pp. 287-302.
62. F. W. McLafferty,(ed.), "Tandem Mass Spectrometry", Wiley, New York, 1983.
63. R. L. Cooks and G. L. Glish, Chem. & Engineering News, Nov. 30, 1980, p. 40.
64. J. C. Kleingeld, Ph.D. thesis, University of Amsterdam (1984).
65. D. L. Bricker, T. A. Adams, Jr., and D. H. Russell, Anal. Chem. (1983) **55**, 2417.
66. W. T. Huntress, M. M. Moseman, and D. D. Ellerman, J. Chem. Phys. (1971) **54**, 843.
67. R. B. Cody and B. S. Freiser, Anal. Chem. (1982) **54**, 1431.
68. R. L. White and C. L. Wilkins, Anal. Chem. (1982) **54**, 2211.
69. M. B. Wise, Anal. Chem., in press.
70. B. S. Freiser, Anal. Chim. Acta (1985) **178**, 137.
71. C. H. Watson, G. Baykut, M. A. Battiste, and J. R. Eyler, Anal. Chim. Acta (1985) **178**, 125.
72. W. D. Bowers, S. S. Delbart, R. L. Hunter, and R. T. McIver, Jr., J. Am. Chem. Soc. (1984) **106**, 7288.
73. C. H. Watson, G. Baykut, and J. R. Eyler, this volume.
74. R. L. Hettich and B. S. Freiser, this volume.
75. F. W. McLafferty, I. J. Amster, J. J. P. Furlong, J. A. Loo, B. H. Wang, and E. R. Williams, this volume.

76. R. B. Cody, Jr. and B. S. Freiser, Anal. Chem. (1987) **59**, 1054.
77. M. T. Bowers, (ed.), "Gas Phase Ion Chemistry", Vols. 1 and 2 (1979) and Vol. 3 (1984), Academic Press, New York.
78. H. B. Linden, H. Knoll, and K. P. Wanczek, paper presented at the 35th Annual Conference on Mass Spectrometry and Allied Topics, Denver, CO, May 24-29, 1987.
79. M. A. Posthumus, P. G. Kistemaker, H. L. C. Meuzelaar, and M C. Ten Noever deBrauw, Anal. Chem. (1978) **50**, 985.
80. R. S. Brown and C. L. Wilkins, this volume.
81. D. B. Jacobson, G. C. Byrd, and B. S. Freiser, J. Am. Chem. Soc. (1982) **104**, 2321.
82. G. D. Byrd and B. S. Freiser, J. Am. Chem. Soc. (1982) **104**, 5944.
83. W. E. Reents, Jr., A. M. Mujsce, W. E. Bondybey, and M. L. Mandrich, J. Chem. Phys. (1987) **86**, 5568.
84. C. L. Wilkins, D. A. Weil, C. L. C. Yang, and C. F. Ijames, Anal. Chem. (1985) **57**, 520.
85. E. C. Brown, P. Kovacic, C. A. Wilkie, R. B. Cody, Jr., and J. A. Kinsinger, J. Polym. Sci., Polym. Lett. Ed. (1985) **23**, 435.
86. D. A. McCrery and M. L. Gross, Anal. Chim. Acta (1985) **178**, 105.
87. R. S. Brown, D. A. Weil, and C. L. Wilkins, Macromolecules (1986) **19**, 1255.
88. D. A. McCrery and M. L. Gross, Anal. Chim. Acta (1985) **178**, 91.
89. R. B. Cody, Jr., I. J. Amster, and F. W. McLafferty, Proc. Natl. Acad. Sci. USA, (1985) **82**, 6367.
90. R. L. Hettich and M. V. Buchanan, paper presented at the 35th Annual Conference on Mass Spectrometry and Allied Topics, Denver, CO, May 24-28, 1987.
91. R. E. Shomo, II, A. G. Marshall, and C. R. Weisenburger, Anal. Chem. (1985) **57**, 2940.
92. M. E. Castro, L. M. Mallis, and D. H. Russell, J. Am. Chem. Soc. (1985) **107**, 5652.
93. D. H. Russell, Mass Spectrom. Rev. (1986), **5**, 167.
94. D. F. Hunt, J. Shabanowitz, J. R. Yates, III, N-Z Zhu, D. H. Russell, and M. E. Castro, Proc. Natl. Acad. Sci. USA (1987) **84**, 620.
95. R. D. Macfarlane, Anal. Chem. (1983) **55**, 1247A.
96. J. C. Tabet, J. Rapin, M. Poreti, and T. Gaumann, Chimia, (1986) **40**, 169.
97. S. K. Viswanadham, D. M. Hercules, R. R. Weller, and C. S. Giam, Biomed. Environ. Mass Spectrom. (1987) **14**, 43.
98. K. Biemann, Anal. Chem. (1986) **58**, 1289A.
99. E. Lippma, R. Pikver, E. Suurmaa, J. Past, J. Puskar, I. Koppel, and A. Tammik, Phys. Rev. Lett. (1985) **54**, 285.
100. M. Bamberg and K. P. Wanczek, paper presented at the 35th Annual Conference on Mass Spectrometry and Allied Topics, Denver, CO, May 24-29, 1987.
101. T. Carlin, "The Fourier Transform Mass Spectrometer as a Gas Phase Chemical Laboratory", Ph. D. Thesis, Purdue University, Aug. 1983.

102. B. S. Freiser, Talanta (1985) **32(8B)**, 697.
103. T. C. Jackson, D. B. Jacobsen, and B. S. Freiser, J. Am. Chem. Soc., (1984) **106**, 1252.
104. A. G. Marshall, T.-C. L. Wang, L. Chen, and T. L. Ricca, this volume.

RECEIVED October 9, 1987

# Chapter 2

# New Excitation and Detection Techniques in Fourier Transform Ion Cyclotron Resonance Mass Spectrometry

**Alan G. Marshall [1,2], Tao-Chin Lin Wang[4], Ling Chen [1], and Tom L. Ricca [3]**

**[1]Department of Chemistry, Ohio State University, Columbus, OH 43210**
**[2]Department of Biochemistry, Ohio State University, Columbus, OH 43210**
**[3]Chemical Instrument Center, Ohio State University, Columbus, OH 43210**

FT/ICR experiments have conventionally been carried out with pulsed or frequency-sweep excitation. Because the cyclotron experiment connects mass to frequency, one can construct ("tailor") any desired frequency-domain excitation pattern by computing its inverse Fourier transform for use as a time-domain waveform. Even better results are obtained when phase-modulation and time-domain apodization are used. Applications include: dynamic range extension via multiple-ion ejection, high-resolution MS/MS, multiple-ion simultaneous monitoring, and flatter excitation power (for isotope-ratio measurements).

The analytically important features of Fourier transform ion cyclotron resonance (FT/ICR) mass spectrometry (1) have recently been reviewed (2-9): ultrahigh mass resolution (>1,000,000 at $m/z \leq 200$) with accurate mass measurement even in gas chromatography/mass spectrometry experiments; sensitive detection of low-volatility samples due to 1,000-fold lower source pressure than in other mass spectrometers; versatile ion sources (electron impact (EI), self-chemical ionization (self-CI), laser desorption (LD), secondary ionization (e.g., $Cs^+$-bombardment), fast atom bombardment (FAB), and plasma desorption (e.g., $^{252}Cf$ fission); trapped-ion capability for study of ion-molecule reaction connectivities, kinetics, equilibria, and energetics; and mass spectrometry/mass spectrometry (MS/MS) with a single mass analyzer and dual collision chamber.

## Ion Motion in Crossed Static Magnetic and Oscillating Electric Fields

The basic principle of FT/ICR mass spectrometry is that a moving ion in an applied static magnetic field undergoes circular motion, in a plane perpendicular to that field, at a "cyclotron" frequency, $\omega$,

[4]Current address: Section on Analytical Biochemistry, Building 10, Room 3D–40, National Institute of Mental Health, 9000 Rockville Pike, Bethesda, MD 20892

0097–6156/87/0359–0021$06.00/0

$$\vec{\omega}_c = q\vec{B}/m \tag{1}$$

in which B is the magnetic field strength (Tesla), q is the ionic charge (Coul), and m is the ionic mass (kg). The cyclotron motion is excited and detected as shown schematically in Figure 1 (10). First, if an electric field oscillating (or rotating) at an angular frequency, $\omega$ (=$2\pi\nu$), in a plane perpendicular to the magnetic field axis, then "resonant" ions (i.e., those for which $\omega = \omega_c$) will be driven coherently forward in their orbits, so that their average ICR radius increases linearly with time. The resonant ions are said to be "excited" by the oscillating (typically radiofrequency) electric field. "Non-resonant" ions (i.e., for which $\omega \neq \omega_c$) will follow a complicated path whose long-term effect is to leave them essentially unmoved from their original positions.

Excited ion orbital radius is theoretically proportional to the product of the magnitude and duration of a resonant rf electric field (11). Moreover, the signal induced in an opposed pair of receiver plates of a trapped-ion cell is proportional to ICR radius (12). The time-domain signal magnitude (or frequency-domain spectral peak area) at a given cyclotron frequency is directly proportional to the number of ions of the corresponding $m/z$ value. Thus, if there is no rf power at the ion's cyclotron resonance frequency (case A in Figure 1), then the incoherent cyclotron motion of a thermal ensemble of ions of common $\omega_c$ and translational energy but random phase (i.e., random angular position in their circular orbit) will induce zero signal in the receiver plates. Ions excited by a resonant rf electric field to an orbit that is large (e.g., 1 cm) but still within the cell boundaries (case B in Figure 1) will be detected with optimal sensitivity. Finally, ions resonantly excited to a radius larger than that of the ICR cell will be lost, or "ejected".

Thus, by appropriate scaling of the rf electric field magnitude at the ICR frequencies of ions of various $m/z$-ratios, we can suppress, excite and then detect, or eject ions at each $m/z$-ratio. The remaining problem is to design the optimal time-domain waveform to accomplish the desired combination of the effects shown at the bottom of Figure 1, over a wide mass range.

## Excitation waveforms

First, it is important to appreciate that the detected ion cyclotron signal magnitude is (to a good approximation) directly proportional to the excitation magnitude, as we have previously demonstrated experimentally (13,14). Thus, in contrast to sector or quadrupole mass spectrometers, it is not necessary to compute the detailed trajectories of the ions in order to determine the effect of those trajectories upon the detector. In other words, we need only characterize the excitation magnitude spectrum (i.e., how much power reaches ions at each ICR frequency) in order to know how much signal to expect at the receiver (for a given number of ions at that cyclotron frequency).

The frequency-domain spectra of four kinds of time-domain excitation waveforms are shown in Figure 2. A simple rectangular radiofrequency pulse was the waveform used to produce the very first FT/ICR signals (1), and is still useful for single-frequency ejection of ions having a narrow range of $m/z$-values. However, the width of its corresponding frequency-domain "sinc" function (Fig.2, top right)

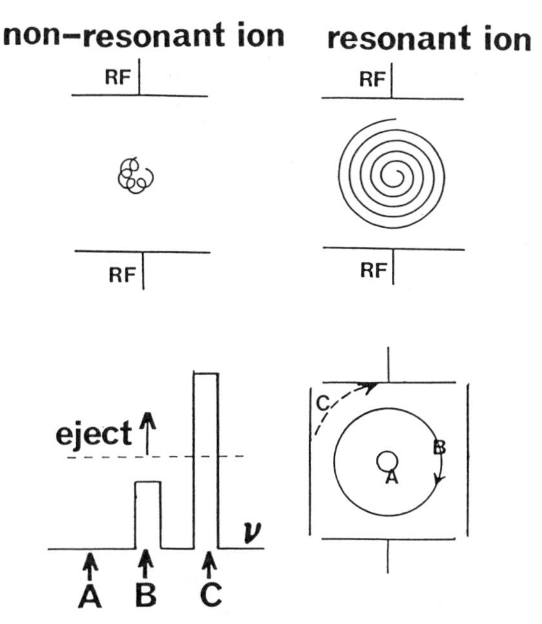


Figure 1. Fundamentals of ICR excitation. The applied magnetic field direction is perpendicular to the page, and a sinusoidally oscillating radiofrequency electric field is applied to two opposed plates (see upper diagrams). Ions with cyclotron frequency equal to ("resonant" with) that of the applied rf electric field will be excited spirally outward (top right), whereas "off-resonant" ions of other mass-to-charge ratio (and thus other cyclotron frequencies) are excited non-coherently and are left with almost no net displacement after many cycles (top left). After the excitation period (lower diagrams), the final ICR orbital radius is proportional to the amplitude of the rf electric field during the excitation period, to leave ions undetected (A), excited to a detectable orbital radius (B), or ejected (C).

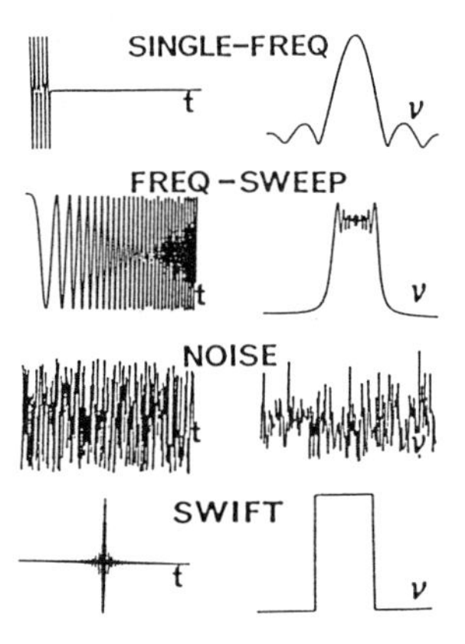


Figure 2. Time-domain excitation waveforms (left) and corresponding frequency-domain magnitude-mode spectra (right) of four excitation waveforms used in FT/ICR. A time-domain rectangular rf pulse gives a "sinc" excitation spectrum in the frequency-domain. A time-domain frequency-sweep gives a complex profile described by Fresnel integrals. Single-scan time-domain noise gives noise in the frequency-domain. Finally, Stored Waveform Inverse Fourier Transform (SWIFT) excitation can provide an optimally flat excitation spectrum (see Figure 3 for details).

varies inversely with the duration of the rf pulse. As a result, pulsed rf excitation is unsuitable for the broadband excitation needed to span a typical mass spectral range (say, 18 < $\underline{m}/\underline{z}$ < 600): an rf pulse of very short duration (<0.1 μs), and thus an impractically high magnitude (>10,000 V/cm) would be needed (11).

Therefore, the second (15) and virtually all subsequent FT/ICR experiments have been performed via frequency-sweep excitation, in which the time-domain rf waveform is swept linearly across the frequency bandwidth corresponding the the $\underline{m}/\underline{z}$-range of interest. Although successful in exciting ICR signals over a wide mass range at relatively low excitation magnitude (ca. 10 V(p-p)/cm), the problems with frequency-swept excitation are evident in Figure 2. First, the excitation magnitude varies significantly (up to ca. ±25% for typical bandwidth and sweep rate) over the mass range of interest. Thus, equally abundant ions of different $\underline{m}/\underline{z}$-ratio can exhibit different apparent intensities (11), and isotope-ratio measurements cannot easily be quantitated.

One could propose pseudorandom (16) or random (17) noise as an excitation source. However, a single time-domain noise waveform produces unacceptably large magnitude variation in the frequency-domain (see Figure 2). The frequency-domain envelope for random noise can be smoothed by frequency-domain averaging of sufficiently large number of scans, but the time required (ca. 10 hr) would be unacceptably long. Flat power can also be extracted from decoding the results of a series of linearly independent pseudorandom sequences (e.g., Hadamard sequences (18)), but again a large number of scans is required.

The best approach, adapted from an earlier proposal by Tomlinson and Hill (19), is to specify the desired frequency-domain excitation profile in advance, and then syntheize its corresponding time-domain representation directly via inverse Fourier transformation. The result of the Tomlinson and Hill procedure is shown at the bottom of Figure 2, in which a perfectly flat, perfectly selective frequency-domain excitation is produced by the time-domain waveform obtained via inverse Fourier transformation of the desired spectrum.

## Stored Waveform Inverse Fourier Transform (SWIFT) Excitation

Unfortunately, the result shown in Figure 2 corresponds to time-shared (i.e., essentially simultaneous) excitation/detection, so that the (discrete) frequencies sampled by the detector are the same as those initially specified in synthesis of the time-domain transmitter signal. However, FT/ICR is more easily conducted with temporally separated excitation and detection periods. In practical terms, the result is that for FT/ICR, we need to know the excitation magnitude spectrum at all frequencies, not just those (equally-spaced) discrete frequencies that defined the desired excitation spectrum.

The left-hand column of Figure 3 illustrates the basic SWIFT procedure. A constant-phase (see below) excitation spectrum is first defined at N equally-spaced discrete frequencies. Inverse Fourier transformation then generates the corresponding digitized time-domain waveform, which may be translated, point by point at a constant rate, by a digital-to-analog converter, amplified, and applied to the transmitter plates of an ICR trapped-ion cell. The time-domain signal is then padded with N zeroes (18,20) before Fourier transformation to give the magnitude-mode spectrum shown at the

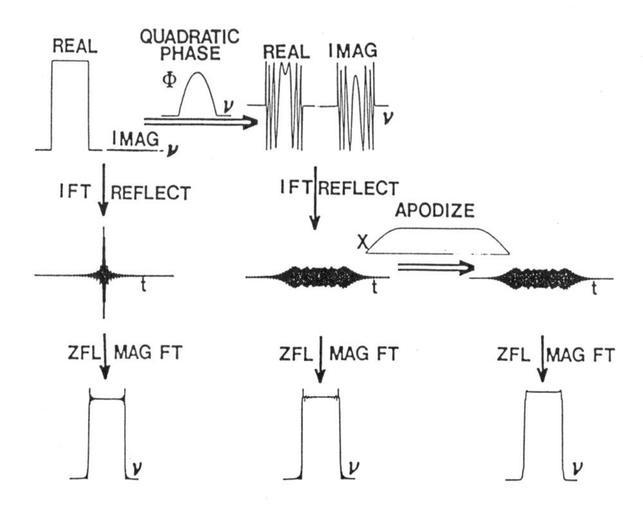


Figure 3. Digital synthesis of a Stored Waveform Inverse Fourier Transform (SWIFT) excitation. Although direct inverse discrete Fourier transform of a constant-phase frequency-domain spectrum (top left) gives a time-domain waveform (middle left) with good frequency-domain uniformity and selectivity (bottom left), the dynamic range of that time-domain waveform can be inconveniently large. Therefore, the original excitation spectrum is first phase-encoded (in this case, quadratically) to give the in-phase and 90°-out-of-phase spectra specified by their magnitude and phase at each discrete frequency (top middle). Inverse Fourier transformation then gives a time-domain waveform (center) with reduced dynamic range and the same final excitation magnitude spectrum (bottom middle). Finally, time-domain apodization (middle right) further smoothes the final excitation magnitude spectrum (bottom right). The time-domain waveforms (middle row) were zero-filled before Fourier transformation, to define the frequency-domain spectral values both at and between the initially specified discrete frequencies of the spectra in the top row of the Figure.

bottom left of Figure 3. Zero-filling effectively interpolates the final frequency-domain spectrum to reveal its values at frequencies equally spaced between the initially specified frequencies. The non-rectangular shape of the final spectrum is simply a consequence of the classical uncertainty principle: namely, that one cannot measure (or specify) a frequency more accurately than about $\pm(1/T)$ Hz via a time-domain waveform of duration, T.

Time-domain waveforms generated from a constant-phase initial spectrum by the SWIFT procedure just described have been successfully applied to broad-band excitation, windowed excitation, and multiple-ion monitoring (11), and to multiple-ion ejection for enhancement of FT/ICR dynamic range (21) as described below. However, three practical problems are associated with the relatively large time-domain vertical dynamic range (see middle left in Figure 3) that can result from that procedure. First, the time-domain dynamic range may exceed the linear response range of the analog transmitter circuitry. Second, the time-domain dynamic range may exceed the digital dynamic range of the digital-to-analog converter (e.g., 12-bit) or even the host computer (e.g., 20 bit) in which the time-domain waveform must be stored. Third, because most of the transmitter power is concentrated into such a short burst, an rf amplifier with much more power (factor of up to 400 compared to frequency-sweep excitation) may be needed.

## Phase Modulation

As discussed in detail in ref. 22, the origin of the large time-domain dynamic range produced by constant-phase SWIFT lies in the common phase relationship between all of the specified frequency-domain spectral components, much as for the centerburst in an FT/IR interferogram or the large initial signal in an NMR free-induction decay (23). The time-domain dynamic range may therefore be reduced by somehow scrambling the relative phases of the initially specified frequency components, as shown in the top row of Figure 3. Phrased another way, what is needed is to distribute the excitation spectrum between its "real" (i.e., in-phase) and "imaginary" (90°-out-of-phase) components, rather than putting all of the signal into the "real" spectrum.

A simple phase-scrambling scheme is a quadratic variation of phase with frequency (see Figure 3, middle column), in which one must be careful not to violate the Nyquist criterion: i.e., the rate of change of phase with frequency, $d\phi/d\nu$, may not vary by more than $\sim$180° from one discrete frequency to the next in the initial spectrum. Subject to that constraint, a larger phase variation, $d\phi/d\nu$, will produce a smaller time-domain dynamic range. The wider the frequency bandwidth for a given time-domain excitation period, the greater will be the potential reduction in time-domain dyanmic range resulting from phase-encoding.

## Apodization

Although phase-encoding solves the problems associated with large time-domain dyanmic range in SWIFT excitation, the final frequency-domain spectrum still exhibits noticeable oscillation in magnitude at frequencies near the limits of the bandwidth. The final spectrum can be smoothed significantly by prior weighting ("apodization") of the stored time-domain waveform before it is sent to the transmitter. Any of several weighting functions designed to bring the time-domain signal smoothly to zero at both ends of its period is suitable. The result for one such apodization function is shown in the right-hand column of Figure 3. The final frequency-domain excitation magnitude spectrum is considerably smoother than without apodization.

### General applications

Figure 4 shows ideal and actual excitation magnitude-mode spectra for representative mass spectral applications. For example, quantitation of relative numbers of ions of different *m/z*-ratios requires ideally flat excitation power over the mass range of interest. The leftmost segments of the spectra of Figure 4 show that stored-waveform excitation offers much flatter power for such applications (e.g., isotope-ratio measurements, gas-phase ion-molecule equilibrium constants, etc.).

Truly simultaneous multiple-ion monitoring (rightmost segments of the spectra in Figure 4) can be performed via ICR. Unlike single-channel (e.g., sector or quadrupole) mass spectrometers, in which ions of only a narrow range of *m/z*-values can be detected at a given instant, FT/ICR offers inherently broadband detection, so that all ions in a given mass range may be detected simultaneously. Thus, by exciting only the frequencies corresponding to the *m/z*-values of interest, FT/ICR automatically provides simultaneous multiple-ion monitoring. In the (typical) situation shown in Figure 4, SWIFT is much more selective than a series of frequency-sweeps because the SWIFT time-domain power is left on for the entire time-domain excitation period, whereas the frequency-sweep power must be turned on and off once for each mass range to be monitored.

### Dynamic range extension

As discussed in detail in ref. 21, a major problem in FT/ICR is the limited mass spectral dynamic range (factor of ~1000:1). Briefly, it is not easy to detect fewer than ~100 ions at easily managed chamber pressures ($\geq 10^{-9}$ torr); alternatively, the spectrum can be distorted by Coulomb forces when more than ~100,000 ions are present.

Consequences of the dynamic range limit are well-illustrated in Figure 5. The normal broad-band FT/ICR mass spectrum of perfluoro-tri-n-butylamine exhibits several strong peaks. Simple expansion of the vertical scale renders some of the small peaks visible, but their presence and quantitation is made difficult by the Coulomb broadening and overlap from abundant ions of nearby *m/z*-ratios.

An obvious solution is to eject selectively all of the most abundant ions, leaving only the small peaks of interest. As argued from Figure 4, multiple frequency-sweeps are not sufficiently selective for this purpose. However, SWIFT tailored ejection of the most abundant ions (23 in this case) removes the broadening and overlap from the large peaks. The resultant spectrum of the less abundant ions (bottom of Figure 5) exhibits better mass resolution (because Coulomb broadening has been reduced) and flatter baseline

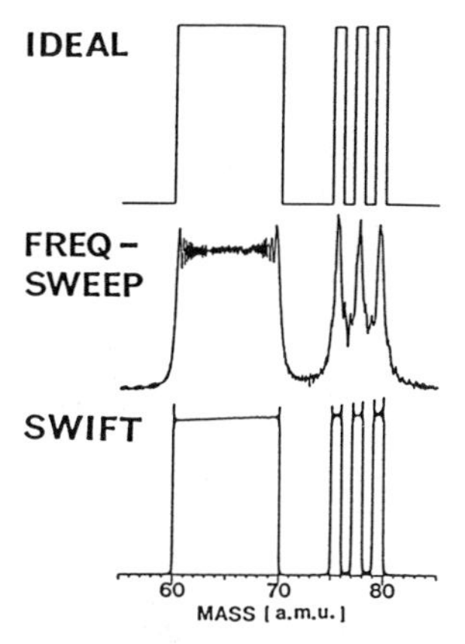


Figure 4. Comparison of a theoretical magnitude-mode excitation spectrum (top) with those detected (on one pair of cell transverse plates) during transmission (on the other pair of cell transverse plates) of a frequency-sweep (middle) or SWIFT (bottom) waveform. The time-domain signals were zero-filled once before Fourier transformation to reveal the full shape of the excitation magnitude spectrum. Note the much improved uniformity and selectivity for SWIFT compared to frequency-sweep excitation.

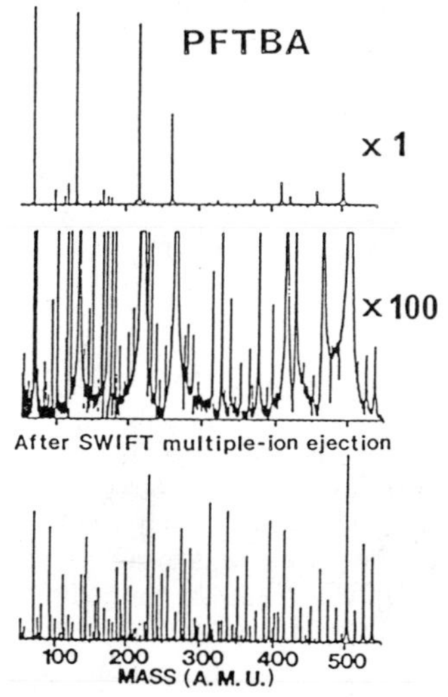


Figure 5. Dynamic range enhancement via SWIFT multiple-ion ejection. Top: Normal broad-band heterodyne-mode FT/ICR mass spectrum of perfluorotri-n-butylamine. Middle: Same spectrum, in which the vertical scale has been expanded such that the peak at $m/z$ = 503 is full scale. Bottom: FT/ICR mass spectrum of the same sample, obtained following prior SWIFT multiple-ion ejection of the (23) peaks whose magnitude-mode peak heights exceeded a threshold of 1.6 % of the height of the biggest peak in the original spectrum. Note the higher mass resolution, reduced noise, and flatter baseline in the SWIFT multiple-ion ejected spectrum.

(because the large peaks (and their broad shoulders) have been eliminated). Moreover, because most of the originally formed ions are removed, it is possible to increase the receiver gain and/or increase the number of initially formed ions to increase dynamic range further.

## High-resolution MS/MS

The potentially ultrahigh mass resolution of FT/ICR detection was one of its earliest appreciated advantages (24). Moreover, the potential of FT/ICR for MS/MS with a single mass spectrometer has been exploited extensively (see especially refs. 5 and 8, citing examples up to MS/MS/MS/MS/MS). In MS/MS via FT/ICR, a typical event sequence might be the selective ejection of all but one ion (i.e., the first "MS"), followed by collision-induced (collision-activated) dissociation in the presence of a steady or pulsed collision gas, followed by broad-band excitation and detection of the mass spectrum of daughter ions (i.e., the second "MS). Although mass resolution in the second "MS" is detector-limited (and therefore potentially ultrahigh), mass resolution in the first "MS" has been limited by the poor selectivity of frequency-sweep excitation (e.g., Figure 4).

The introduction of the SWIFT technique (10,14,21,22) makes possible FT/ICR frequency-domain excitation with the same mass resolution as has already been demonstrated for FT/ICR detection, provided only that sufficient computer memory is available to store a sufficiently long time-domain waveform. When ejection must be performed with ultrahigh mass resolution over a wide mass range, a simple solution is to use two successive SWIFT waveforms: first, a broad-band low-resolution excitation designed to eject ions except over (say) a 1 amu mass range; and then a second SWIFT waveform, heterodyned to put $\geq$ 8K data points spanning a mass range of 1-2 amu.

Finally, SWIFT excitation also offers a way to compress the sequence even further by combining the ejection and excitation stages. As shown in Figure 6 for two isotopic ions of nominal mass 92, from electron ionization of toluene at $0.5 \times 10^{-9}$ torr, it is possible to eject the more abundant ion while exciting its narrowly resolved neighbor. Each excitation spectrum was defined by quadratically modulated real (8K) imaginary (8K) frequency-domain points, to give a 16K time-domain stored data set, transmitted over a period of 65.920 ms to give a 125 kHz bandwidth centered at a radiofrequency carrier frequency of 500 kHz (10).

## Conclusion

Stored Waveform Inverse Fourier Transform (SWIFT) excitation for FT/ICR is a newly implemented technique which includes all other excitation waveforms as subsets. Compared to prior excitation waveforms (e.g., frequency-sweep), SWIFT offers flatter power with greater mass resolution and the possibility of magnitude steps (without additional delays or switching transients) in the excitation spectrum. Briefly, SWIFT increases the mass resolution for FT/ICR excitation to the ultrahigh mass resolution already demonstrated for FT/ICR detection.

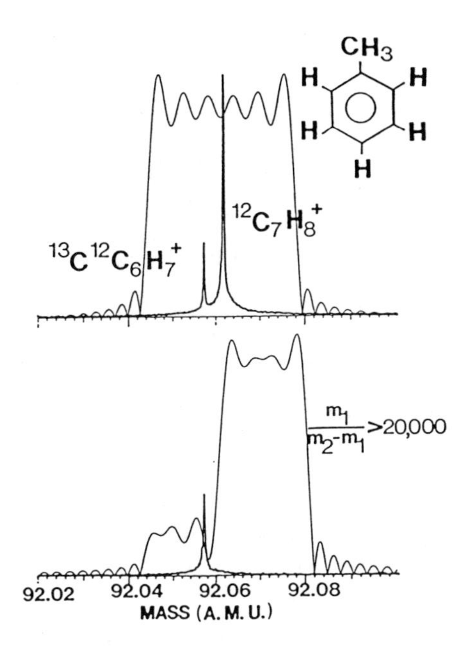


Figure 6. Simultaneous FT/ICR excitation/ejection of two ions of very similar mass-to-charge ratio, produced via electron ionization of toluene. In each plot, the heavy line represents the experimental FT/ICR mass spectrum, and the light line represents the magnitude-mode excitation spectrum used to produce that ICR signal. Top: $^{12}C_6H_8^+$ and $^{13}C^{12}C_6H_7^+$ are excited with uniformly tailored SWIFT excitation extending from $m/z$ = 92.04 to 92.08. Bottom: FT/ICR mass spectrum obtained via SWIFT simultaneous excitation/ ejection, in which $^{13}C^{12}C_6H_7^+$ at mass 92.058 is excited while $^{12}C_6H_8^+$ at mass 92.062 is ejected, at a ratio of 1:5 in excitation magnitude. The ultrahigh selectivity (($m_2$/($m_2$ -$m_1$) >20,000) of such experiments should be valuable for high-resolution MS/MS (see text).

GLOSSARY

m/z: ionic mass-to-charge ratio, in which m is in atomic mass units and z is in multiples of the electronic charge.

Signal amplitude: maximum vertical displacement from equilibrium for a sinusoidally oscillating waveform. Signal amplitude is thus half of the peak-to-peak vertical displacement.

Signal power: square of signal amplitude. For a transverse (e.g., electromagnetic) wave, the power is the energy deposited per unit time per unit cross-sectional area.

Signal magnitude: square root of signal power.

Sinc function: sinc(x) = sin(x)/x. The sinc function is most often encountered as the (frequency-domain) Fourier transform of a (time-domain) rectangular pulse.

Windowed excitation: selective excitation with uniform power over specified frequency range(s), and with zero power over other frequency range(s) (window(s)) for ICR signal suppression.

For other specialized terms of Fourier transform spectroscopy (e.g., interferogram, centerburst, etc.), see ref. 23.

## Acknowledgments

This work was supported by grants (to A.G.M.) from the U.S.A. Public Health Service (N.I.H. 1 R01 GM-31683) and The Ohio State University.

## Literature Cited


1. Comisarow, M. B.; Marshall, A. G. Chem. Phys. Lett. 1974, 25, 282-283.
2. Marshall, A. G. Acc. Chem. Res. 1985, 18, 316-322.
3. Gross, M. L.; Rempel, D. L. Science (Washington, D.C.) 1984, 226, 261-268.
4. Wanczek, K. P. Int. J. Mass Spectrom. Ion Proc. 1984, 60, 11-60.
5. Freiser, B. S. Talanta 1985, 32, 697-708.
6. Laude, D. A., Jr.; Johlman, C. L.; Brown, R. S.; Weil, D. A.; Wilkins, C. L. Mass Spectrom. Rev. 1986, 5, 107-166.
7. Russell D. H. Mass Spectrom. Rev. 1986, 5, 167-189.
8. Nibbering, N. M. M. Comments At. Mol. Phys. 1986, 18, 223-234.
9. Comisarow, M. B. Anal. Chim. Acta 1985, 178, 1-15.
10. Chen, L.; Marshall, A. G. submitted for publication.
11. Comisarow, M. B. J. Chem. Phys. 1978, 69, 4097-4104.
12. Marshall, A. G.; Roe, D. C. J. Chem. Phys. 1980, 73, 1581-1590.
13. Marshall, A. G.; Wang, T.-C. L.; Ricca, T. L. Chem. Phys. Lett. 1984, 105, 233-236.
14. Marshall, A. G.; Wang, T.-C. L.; Ricca, T. L. J. Amer. Chem. Soc. 1985, 107, 7893-7897.
15. Comisarow, M. B.; Marshall, A. G. Chem. Phys. Lett. 1974, 26, 489-490.

16. Ijames, C. F.; Wilkins, C. L. Chem. Phys. Lett. 1984, 108, 58-62.
17. Marshall, A. G.; Wang, T.-C. L.; Ricca, T. L. Chem. Phys. Lett. 1984, 108, 63-66.
18. Marshall, A. G., In "Fourier, Hadamard, and Hilbert Transforms in Chemistry"; Marshall, A. G., Ed.; Plenum: New York, 1982; pp. 1-43.
19. Tomlinson, B. L.; Hill, H. D. W. J. Chem. Phys. 1973, 59, 1775-1784.
20. Comisarow, M. B.; Melka, J. D. Anal. Chem. 1979, 51, 2198-2203.
21. Wang, T.-C. L.; Ricca, T. L.; Marshall, A. G. Anal. Chem. 1986, 58, 2935-2938.
22. Chen, L.; Wang, T.-C. L.; Ricca, T. L.; Marshall, A. G., Anal. Chem. 1987, 59, 449-454.
23. Marshall, A. G. In Physical Methods in Modern Chemical Analysis; Kuwana, T., Ed.; Academic: New York, 1983; pp. 57-135.
24. Comisarow, M. B.; Marshall, A. G. J. Chem. Phys. 1976, 64, 110-119.

RECEIVED June 15, 1987

# Chapter 3

# Problems of Fourier Transform Mass Spectrometry

## A Route to Instrument Improvements

**Richard P. Grese, Don L. Rempel, and Michael L. Gross**[1]

**Midwest Center for Mass Spectrometry, Department of Chemistry, University of Nebraska, Lincoln, NE 68588**

Progress in FTMS is tied to understanding dynamics in ion traps. A three stage approach is envisioned: (1) modeling of ion dynamics, (2) developing tests of models by measuring ion motion in various cells, and (3) implementing new instrumentation such as improved cells and methods of excitation. Elements of this cycle are addressed with respect to five different problems. The first, nonuniform tuning over wide mass ranges, is attributed in part to z-mode excitation. A second problem is temporal variation of an inert ion signal which may lead to inaccurate kinetic results in ion-molecule reaction studies. Systematic errors in mass measurement represent the third problem. Low mass resolution for high mass ions and for ion selection in MS/MS experiments are the fourth and fifth problems that are addressed.

Although there are many attractive features of Fourier transform mass spectrometry (FTMS), there are problems with current implementations. The FT mass spectrometer is still a relatively young instrument when compared to the double-focussing mass spectrometer, and comparisons of the spectrometers and their development histories are informative and encouraging. As the attention paid to problems of earlier single-focussing instruments led to the double-focussing mass spectrometer, for example, investigation of FTMS problems now will lead to better implementations of FTMS. Here, we describe some examples, not all of them complete, of systematic attempts to investigate problems in current implementations of FTMS with an eye on improvement of the instrument.

Comparison with Sector Instruments. Modern double-focussing mass spectrometers have a long history of development, which was recently reviewed (1). The instrument has evolved through a long series of complex and interweaving events beginning with the early experiments of J. J. Thomson (2) and the first mass spectrographs of

[1]Correspondence should be addressed to this author.

0097–6156/87/0359–0034$07.25/0

Dempster in 1918 (3) and Aston in 1919 (4). Early instruments were used to measure atomic masses, and much effort was devoted to increase the resolution of the measurements. To improve resolution, techniques for reduction of the effects of velocity and angular dispersion at neighborhoods of focal points or planes were introduced. For example, the double-focussing instrument first developed by Bartky and Dempster in 1929 was designed to solve problems caused by the energy spread of the ion beam (5). This might be viewed as the introduction of symmetry: ideally we would want an analytical instrument to be invariant to (symmetrical with respect to) all variables except the one being measured. Another example of the use of symmetry in mass spectrometry is the improvement of resolution in time-of-flight instruments by addition of the reflectron. Sector mass spectrometer development still continues; for example, extending the upper range mass limit is an exciting development (6). Techniques employed in recent designs include inhomogeneous magnets, quadrupole lenses, non-normal poleface boundaries, non-ferrous electromagnets (7), and new and more efficient ion sources.

Over sixty years have passed since the beginning of the development of the mass spectrometer as a modern analytical instrument. In comparison, FTMS is a relatively new technique, first introduced by Comisarow and Marshall in 1974 (8). Perhaps it is an indication of the technique's high potential that FTMS has developed so rapidly. However, it should not be surprising, by analogy with the long development of sector instruments, that its full potential has not yet been realized.

Double-focussing instruments have set the standard by which other mass spectrometers, including FT mass spectrometers, are judged. High resolution, parts-per-million mass accuracy, picogram detection limits, and high mass limits (ca. 10,000 u) are attributes shared by both sector and FT mass spectrometers (9). However, achieving high performance is reasonably routine with sector instruments but not yet with FT mass spectrometers.

Resolutions of 500,000 have been reported by Ogata and Matsuda for a sector mass spectrometer (10). Resolving powers of $10^5$ are trivial to obtain even with low field electromagnet-based FT systems and resolutions as high as $10^8$ for an ion of m/z 18 were reported in a high field FTMS (11). Although the resolution for FTMS is higher, there may be an analogy between mass resolution achievements with sector and FT mass spectrometers. For the sector instrument, the resolution was achieved, in part, by making the spectrometer larger. In a sense, this is the same as reducing electric field problems by turning to higher magnetic fields in FTMS, an approach that does not involve the introduction of symmetry.

Because of the respect accorded high resolution sector instruments, the early reports of high resolution with FTMS raised high expectations. It was expected that the higher mass resolution would lead to high mass accuracy. Yet there are no reports of analytical applications in which mass measurement accuracies in FTMS (12) are significantly better than the ppm accuracies achieved for sector machines. Thus, the higher resolution capabilities of FTMS have yet to be translated into mass measurements more accurate than those available from sector machines. Because FTMS is no better, the current accomplishments in mass measurement accuracy are not noteworthy.

The sensitivity of FTMS is also comparable to that of sector instruments. Although sector instrument detectors can be used to count single ions and FTMS detectors cannot, both methods will yield a peak profile for approximately 100 ions. Detecting a single ion in organic or bioanalytical chemistry is of limited utility because its mass cannot be assigned with any certainty.

In addition to the characteristics shared by FTMS and sector MS, FT mass spectrometers have several advantages over sector instruments. These advantages have been cited often and they include (9b): 1) ion trapping and manipulation, 2) multichannel advantage, 3) high resolution with no loss in sensitivity, 4) compatibility with pulsed devices (e.g. lasers, pulse valves), and 5) multiple MS/MS experiments ($MS^n$). Some of these advantages are shared by time-of-flight instruments; however, TOF does not give high resolution or permit accurate mass determinations. Another advantage of FTMS is the approximate mass invariance of the signal intensity (13-14) for a given number of ions and circumstances. This contrasts with the mass discrimination (15) which occurs for multipliers that are used in sector machines.

Status of FTMS. Since its introduction in 1974, FTMS has been demonstrated to be capable of solving a wide variety of chemical problems that are not possible to solve by using other mass spectrometers. For example, analytical applications have been designed to take advantage of the multichannel advantage and the ultra high resolution capabilities of FTMS (16). Ultra high resolution was demonstrated with chemical ionization (17-18), GC/MS (19), multi-photon ionization (MPI) (20), MPI-GC/MS (20), and laser desorption (21). A recent application of FTMS is GC/MS of olefins that are chemically ionized with $Fe^+$ ions which are produced by pulsed laser multiphoton dissociation/ionization of $Fe(CO)_5$ (22). However, much of the research has not addressed fundamental problems in the understanding of the behavior of ions in a cubic cell. Because of our incomplete knowledge of ion dynamics, FTMS remains nonroutine with respect to a number of items. 1) The upper end of the dynamic range is bound by the number of ions the cell can hold. A dynamic range of $10^2$-$10^3$ has been estimated for the cubic cell; this is in contrast to conventional mass spectrometers with electron multipliers which have dynamic ranges of at least $10^5$. 2) Nonuniform tuning over wide mass ranges contributes to unreliable peak heights in the mass spectrum. Correct peak heights are needed for interpretation and for optimum utilization of mass spectra library searches. 3) MS/MS experiments have been limited by poor resolution in MS-I and by poor reproducibility of activation methods. 4) Soft ionization methods have not been routine with FTMS. The use of external sources may be the expedient but not long range solution to this problem.

Route to Improvement. As one route to solving problems that perplex FTMS, we have made use of a systematic approach illustrated in Figure 1. The "Development Cycle" enables us to define clearly problems and to arrive at solutions through an organized combination of theoretical and experimental work. The most difficult aspect of this approach is knowing how to start. An obvious approach is to start with defining the problem; however, multiple causes of problems and mixed symptoms have slowed progress. Once begun, the cyclic nature of this approach grows outward. Improved models provide a basis for design changes that improve instrument performance, and understanding the actual performance leads to more accurate models.

In the remaining sections, specific examples of how this approach has led to instrument improvements are given.

## Solving Problems Facing FTMS

Nonuniform Tuning of Mass Spectra (23). 1. Problem Definition: The motivation for improving our understanding of FTMS develops often when problems arise that limit the accuracy of experimental results. Two such problems: 1) nonuniform tuning for peak heights over wide mass ranges (24), and 2) ion losses in double resonance experiments occurring at excitations insufficient to raise the ion to a radius necessary for ejection (25), have led us to an investigation of z-mode excitation as a possible cause.

Some mechanisms of z-mode excitation are known; and in particular, Beauchamp and Armstrong's (26) direct method of z-excitation can be used for double resonance. Fortunately, excitation of the z-mode via these processes can be avoided by eliminating the excitation frequency components to which the trapping mode respond. This is possible because these frequencies are significantly different than those used to excite the cyclotron mode.

However, excitation of the z-oscillation mode may also occur with frequency components near or at the cyclotron frequency which is essential for excitation of the cyclotron mode. As is exaggerated in Figure 2, when the ions are at the top of the cyclotron orbit, there is a peak positive potential on the top excitation plate that results in a component force toward the trapping plates. Similarly, when the ions are at the bottom of their cyclotron orbit, there is a peak positive potential on the bottom plate resulting also in a component force toward the trapping plate. This synchronization of temporal changes in the z-component of the excitation field and the spatial motion of the ion results in an average force that changes the trapping motion. The energy of the z-mode oscillation might even be modified to the point where the ion is lost from the cell.

2. Model: A simple model (Equation 1) was employed to depict the average force that modifies the z-motion during excitation.

$$F = qE_z \sim qkV_p[z/a][A_r/a](1 - [A_r/a]^2)\cos\phi \qquad (1)$$

$F$ = average force that modifies the z-motion of the ion
$E_z$ = electric field
$V_p$ = amplitude of excitation waveform
$z$ = ion position along the z-axis ($z = 0$ at the cell center).
$a$ = length of inside edge of cubic cell
$A_r$ = radius of cyclotron orbit
$\phi$ = phase of excitation waveform

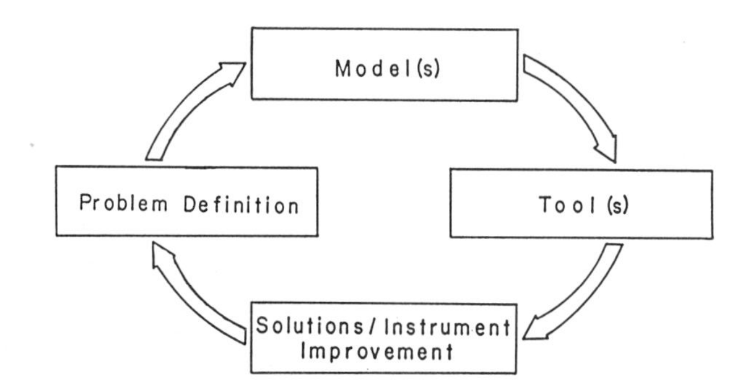


Figure 1. Development cycle for instrument improvements in FTMS.

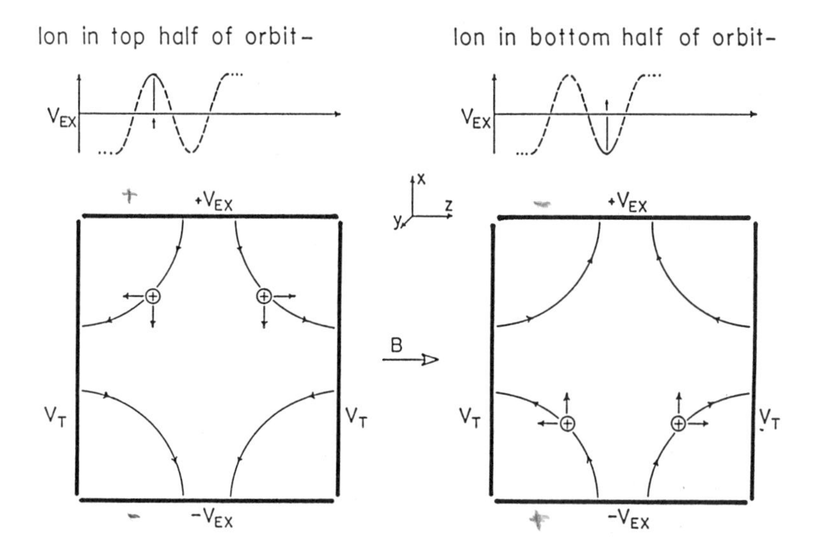


Figure 2. The synchronization between rf excitation waveform and ion cyclotron motion shown with exaggerated phase; i.e. $\phi \neq \pm \pi/2$. (Reproduced with permission from Ref. 23 Copyright 1986 Elsevier Science Publishers B.V.)

Several trends are predictable with this simplified model. 1) The perturbation of the z-motion increases with the excitation amplitude $V_p$. This suggests that the z-oscillation would be largest for the high amplitude rf bursts that are used to achieve greater bandwidths in FTMS. 2) The perturbation becomes smaller as the z-position of the ion goes to zero. This suggests that processes that reduce the magnitude of the z-oscillations before excitation, such as ion-molecule collisions, should reduce the undesired z-excitation when the perturbing force is transitory. 3) The perturbation also depends on the relative phase between the ion motion and the excitation waveform. This dependence suggests a transitory perturbation for chirp excitations that is mass discriminatory.

3. Tool: To demonstrate z-mode excitation, it is easiest to design an experiment that measures the phenomenon as it occurs in the extreme case; that is, ejection of the ion along the z-axis. The "trap-switch" experiment (23) (Figure 3a) determines the ion loss in the z-direction as a function of time by rapidly increasing the trapping voltage at a variable time in the sequence. As the trap is switched to a higher voltage, the threshold for z-mode loss is increased. Changes in signal reflect differences in the ion loss between the low (test) trapping voltage (0.44 or 1.04 V) and the high (reference) voltage (4 V) over the time that the trap is at the lower voltage. If there is z-mode ion loss during excitation, then the ion signal when the trap-switch occurs after excitation should be lower than when the trap-switch occurs prior to excitation. Less ion loss should be seen if a higher trapping field is on during the test excitation.

The first result of the trap-switch experiment (see Figure 3b) is the loss of signal occurring when the trap-switch is after the test chirp excitations. The decrease in signal around the time of the excitation indicates that ion loss occurred in the z-direction as a result of the excitation.

Several additional observations are made: 1) There is little difference in the number of ions lost as a consequence of wide band (0-2000 kHz) chirp excitation for trapping voltages of either 0.44 or 1.04 V at high level excitation in some variance with expectation. 2) Wide band chirps (0-2000 kHz) produce larger ion losses than do narrow band chirps (200-350 kHz), perhaps, because excitation of the magnetron mode occurs in the first case. 3) Ion losses decrease as the amplitude of the wide band chirp is decreased as expected from the simple model. 4) Ion z-ejection is considerably reduced if the test chirp excitation was delayed from 62 ms to 2.015 sec (plotted as open squares in Figure 3b). As expected, collisional damping of the z-mode amplitudes during the long delay has reduced the magnitude of z-mode excitation.

Another tool is the numerical integration of the ion equations of motion. McIver (27) and Marshall (28) have made use of this tool for studying some aspects of ion motion. Here, to corroborate that mass discriminating z-losses do occur, the trajectory of single ions were calculated under conditions chosen to imitate normal conditions for which symptoms of z-excitation are observed. For the calculations, the ions were started on the z-axis and were given different z-mode energies. The chirp excitation was 9.67 volts base-to-peak at each excitation plate with a 2.105 kHz/µsec sweep

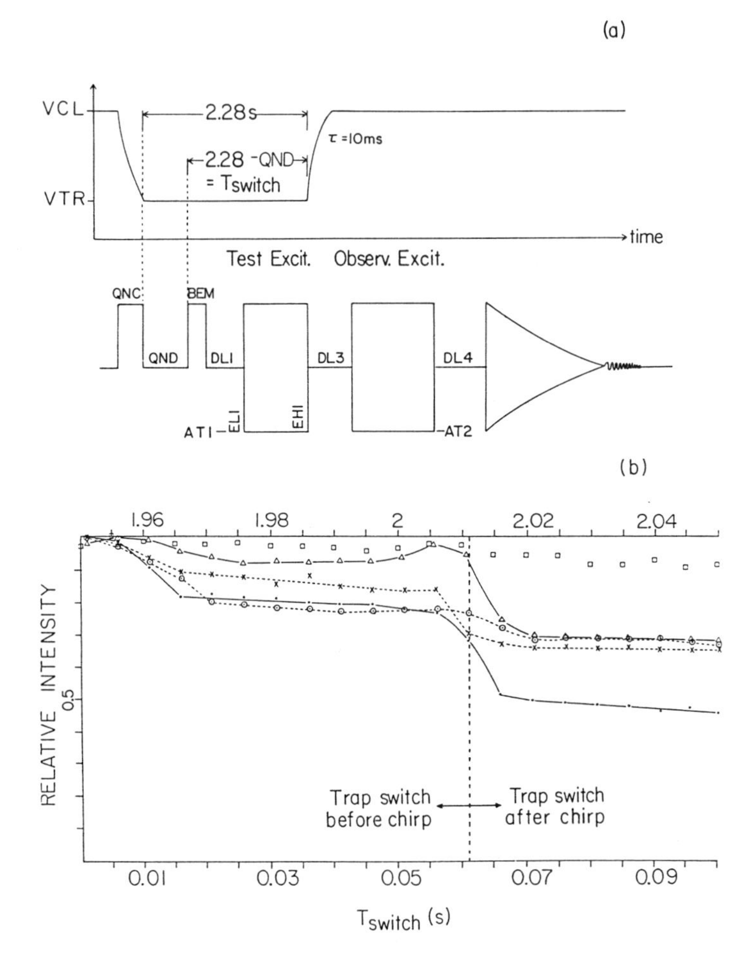


Figure 3. (a). Trap switch time sequence. (b). Relative intensity of benzene molecular ion versus trap switch time for a chirp test excitation. 100,000 ions were formed initially.

| | EL1(kHz) | EH1(MHz) | AT1(dB) | VTR(V) | DL1(ms) |
|---|---|---|---|---|---|
| •---• | 0 | 2 | 2 | 0.44 | 47 |
| △---△ | 0 | 2 | 2 | 1.04 | 47 |
| ○---○ | 0 | 2 | 5 | 0.44 | 47 |
| x...x | 200 | 0.350 | 2 | 0.44 | 47 |
| □...□ | 0 | 2 | 2 | 0.44 | 2000 |

(Reproduced with permission from Ref. 23. Copyright 1986 Elsevier Science Publishers B.V.)

rate and 0 to 2 MHz sweep width. Other conditions were a 1.2 T magnetic field, a 0.0254 m cubic trap, and a 1.0 V trap potential.

When the z-mode position and velocity at the beginning of the excitation is compared with the corresponding information at the conclusion of the excitation, it is found that in all the cases tried for benzene (m/q = 78) and nitrogen (m/q = 28), the z-mode energy is increased by the excitation. Those benzene ions with initial ion z-mode energies 50 percent of the trappable maximum are lost in the z-direction whereas ions with initial z-modes of 40 percent or less are not lost. Those nitrogen ions with initial z-mode energies 17.5% or more are lost whereas an ion with an initial z-mode energy of 15% is not lost. The different initial z-mode energies required for loss is consistent with the hypothesis that for lower mass ions a larger fraction of the ion population is ejected.

The numerical results are consistent with the result of an experiment in which the ion cyclotron orbit sizes of a methane ($CH_4^+$) and benzene ($C_6H_6^+$) mixture of ions were varied. In the control experiment, the two ions were excited by low amplitude consecutive RF burst pulses of varied time. The signal ratio was essentially constant over the range of orbits for which signals were detectable. In contrast, for a chirp from 10 kHz to 2 Mhz at 2.094 kHz/μsec of varied amplitude, the abundance ratio of $CH_4^+/C_6H_6^+$ decreased from about 90% to about 10% as the orbit size was increased, indicating loss of the lighter ion.

4. Possible Solutions: Because z-mode excitations degrade the performance of FTMS, methods to minimize z-mode excitations are necessary. Theoretical (Equation 1) and experimental results suggest several possible strategies.

i) Low amplitude rf bursts should be used to select specific ions when quantitation is the goal and wideband operation is not essential.

ii) Excitation pulses might be used if they are short enough to impart the momentum to the ions before they have a chance to move a significant distance from the z-axis.

iii) The ions should be confined to smaller z-mode amplitudes before exciting the cyclotron mode. Two methods to facilitate this involve using the trap-switch to compress the ion cloud prior to excitation or inserting sufficient delay time between ionization and excitation to allow collisional relaxation of the ion cloud.

iv) Modified cell designs might be used to minimize z-loss. a) Elongating the cell in the z-direction (29) should reduce the z-component electric forces and decrease the fraction of ions exposed to the larger z-components of the excitation field. b) Modification of the cell design to remove the odd symmetries of the cubic cell that cause the synchronization should also be useful. Two such modified cells are the cylindrical (30) and hyperbolic traps (31). c) Formed excitation and detection plates may constitute another cell modification for reducing the effects of z-mode excitation by improving the uniformity of the excitation fields. Formed cell plates were used by Clow and Futrell to improve the uniformity of drift velocities in a drift cell ICR (32). Similarly, for cubic cells the excitation field uniformity would be improved by bringing all the edges of two nominally parallel plates closer together to compensate for their finite area and other electrode proximity (Figure 4). The same set of plates would be used for both excitation and detection by using carefully designed switches and in that way escape the dilemma of designing one set of plates at the expense of the other.

Temporal Variation of Ion Signal of an Inert Ion. 1. Problem Definition: Several assumptions are required if correct correlations between ion signal and chemical reactivity are to be made from ion kinetic studies using FTMS. The first assumption must be that the sensitivity of the ion trap remains constant with time so that ion signals are representative of the ion population with time. The second assumption is that losses of reactant ions are due to consumption of those ions in a chemical reaction and not by physical processes.

One would expect that the signal of an unreactive ion would be invariant with time. However, the benzene molecular ion signal decreases or increases in time for correspondingly low or high trap voltages. Decreases in ion abundance are not surprising, as ion radial diffusion will occur; but the ion abundance increases shown in Figure 5 are more unexpected.

2. Model: Ion losses from an FTMS cell can occur by mechanisms other than chemical reaction and radial diffusion. Ions that have acquired sufficient kinetic energy from fragmentation processes can overcome the potential barrier of the trapping field especially when low trapping voltages are used. Ion loss owing to fragmentation was demonstrated by Riggin, et al. (34a,b).

Ion evaporation can also account for ion losses from the cell. The term "evaporation" is used to describe processes by which ion-ion collisions distribute z-mode energy with the result that some ions acquire enough kinetic energy to escape the trapping field. Ion evaporation was studied for thermal population of ions at low pressure ($3x10^{-10}$ torr) and long delay times (100 sec) (35). Ion evaporation can be reduced with higher trap voltages that provide higher thresholds for loss. As shown in Figure 5 for the case of the benzene molecular ion, further increases in trap voltage beyond 0.74 volts produce relatively small further increases of signal with time.

After reducing the magnitude of ion evaporation, we are still left with a signal that increases in time. In this regard, ion image currents have been shown to be dependent on the position of ions cyclotroning with small orbits in rectangular traps (36). Because ions change their positions in the cell as a function of time, the signal per ion may be expected to change. Furthermore, signals from ions formed by ionization methods that are directed along axes other than the z-axis (such as laser desorption and multiphoton ionization in the cell we are using) may be affected by the different spatial distribution of the ions.

We have developed a model to explain the time dependent change in sensitivity for ions during excitation and detection. The first step is to describe the image charge displacement amplitude, $S(A_R, A_z)$, as a function of cyclotron and z-mode amplitudes. The displacement amplitude was derived using an approximate description of the antenna fields in a cubic cell. The second step in developing the model is to derive a relationship to describe the cyclotron orbit as a function of time for an rf burst. An energy conservation

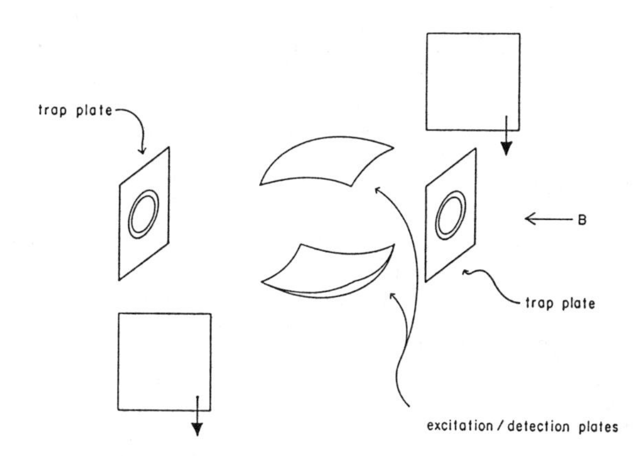


Figure 4. Exploded modified cubic trap with formed excitation/detection plates. Trap electrodes are segmented by electrically isolating a disk shaped region.

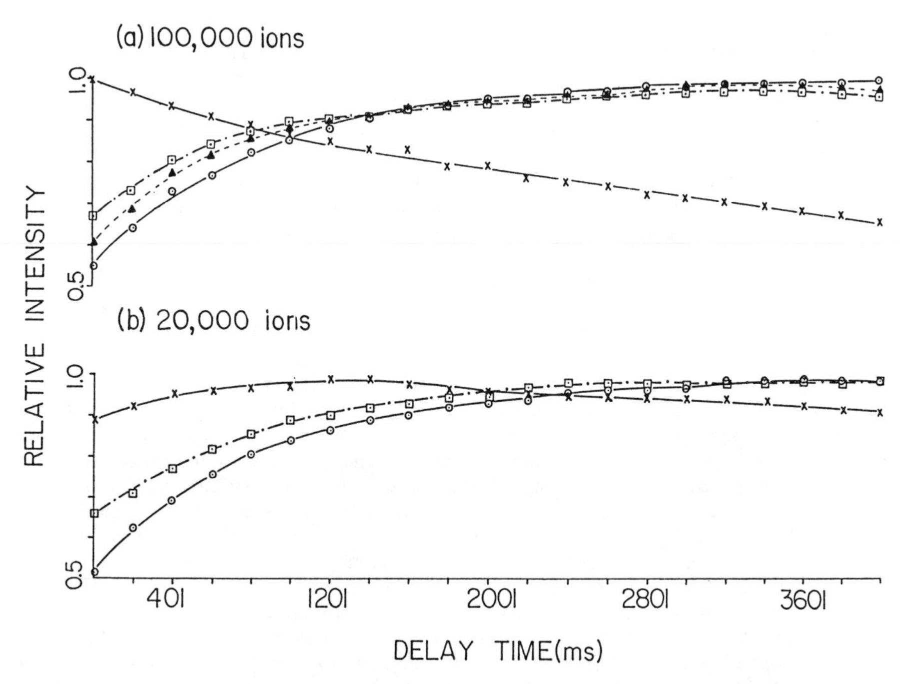


Figure 5. (a). Temporal dependence of relative signal intensity of 100,000 benzene ions at different trap voltages: x, 0.44 V; ⊡, 0.74 V; ▲, 1.04 V and ⊙, 2.04 V. (b). Initial 20,000 benzene ions at trap voltages of x, 0.44 V; ⊡, 0.74 V; and ⊙, 2.04 V. (Reproduced with permission from Ref. 33. Copyright 1986 Elsevier Science Publishers B.V.)

argument is used to deduce the cyclotron orbit sizes from the charge displacement on the excitation plates. Because the symmetry of the cubic cell requires that the charge displacement amplitude on the receiver plates be the same as on the excitation plates, the amplitude of the receiver charge displacement can also be inferred. By using the above arguments applied to a narrow band rf excitation, a relationship between $A_r$, the radius of the cyclotron mode, and $A_z$, the z-mode oscillation amplitude, can be obtained. This relationship (see Figure 6) is then used to deduce the signal, $S(A_z)$, as a function of $A_z$ only.

The most obvious effect seen in Figure 6 is the large dependence of ion signal on ion z-oscillation amplitudes. The signal, normalized to 100% for $A_z = 0$, drops to <30% for ions grazing the trap plates. It is also apparent from Figure 6 that for ions of low $A_z$ values, the graph is fairly flat. Thus, if the z-mode oscillation amplitude of each ion is initially confined to a low value, the signal change owing to further collisional damping of the z-mode amplitude, $A_z$, should be reduced. The figure also shows a weak dependence of curve shape on the size of $A_r$.

It is also expected that reactive collisions may diminish the effects of collisional damping of the z-oscillation. An unreactive collision removes energy from the z-mode oscillation so that the ion contributes more signal current at its original cyclotron frequency whereas a reactive collision removes an ion from a reactant population giving a true indication of the loss from the original population. The loss rate from the reactant population for ions of z-oscillation, $A_z$, is proportional to the density of reactant ions of amplitude $A_z$. Thus, for very reactive ions, no change in sensitivity due to collisional relaxation is expected.

3. Tool: According to the above model, the signal for an unreactive ion should increase as the ion cloud undergoes relaxation toward the z = 0 plane. The size of this change can be computed as a ratio of expected signals for two z-mode amplitude distributions. The first is the average of $S(A_z)$ over a z-mode amplitude distribution determined by uniform initial positions along the z axis and a thermal distribution of z velocities corresponding to the temperature of the background gas. This approximates the z-mode amplitude distribution of the benzene molecular ions generated by an electron beam assuming no collisions occur after the ion is formed. The second is the average of $S(A_z)$ over a thermal z-mode amplitude distribution determined by a quadratic potential along the z-axis and with a temperature equal to that of the background gas. Here it is assumed that ion-neutral collisions dominate to determine the second distribution.

The tool to investigate the relationship between sensitivity and z-mode relaxation is simply a measurement of the signal intensity of an unreactive species as a function of time under conditions in which evaporation and charge exchange are known to be small effects. The measured ratio of signal (just after the beam) to signal maximum (after a delay of seconds) compares reasonably well with the ratio computed by employing the assumed distributions.

4. Possible Solutions: For accurate kinetic measurements, the effect of z-mode relaxation on ion signals must be controlled. It is impractical to use long delay times between ionization and

excitation because time delay is the experimental variable. A possible solution is ion compression using a trap-switch (23). Low kinetic energy ions should be compressed to ca. the z = 0 plane by increasing the trapping potential after ionization. Figure 7 shows the effect of ion compression on the temporal behavior of the benzene molecular ion signal. The effects of z-mode relaxation at short delay times have been largely removed by the use of the trap-switch.

Systematic Errors in Accurate Mass Measurements. 1. Problem Definition: The value of high resolution mass spectrometry is diminished if the mass measurements do not give unambiguous elemental compositions. Accurate mass measurements in FTMS require a precise measurement of ion frequencies and an accurate calibration law for converting ions frequencies to mass. The ion frequencies can be measured to nine significant figures with modern electronics; however, the relationship between ion frequencies in the cubic cell and mass still requires further development.

The evolution of a calibration law for the cubic cell has seen a number of proposals for the frequency-to-mass relationship (26,37-40). Recently, Ledford, *et al.* (40) developed an algebraically correct mass calibration law (Equation 2), which is now commonly used. Unlike sector mass spectrometry in which many

$$m = a/f_{obs} + b/f^2_{obs} \tag{2}$$

reference mass ions are required, only two references are required to calculate the two calibration constants, (2) a and b, which are required to relate mass, m, to the observed signal frequency, $f_{obs}$. This equation is sufficiently accurate to give errors in the 1-10 ppm range (12). However, in some instances, these errors are systematic and too large to give correct elemental composition assignments. Also, the systematic errors have been observed to increase with increasing number of ions. Therefore, the model does not account for all space charge effects. Nevertheless, it has been estimated that 70% of the space charge effects are accounted for (40). Space charge-related errors may limit utility of the calibration law over wide dynamic ranges of ion numbers such as are encountered in GC/FTMS.

2. Model: A quantitative model that leads to the development of an accurate calibration law is not yet readily obtainable because of the apparent complexity in the frequency variations in the cubic cell. Frequency variations have been studied by Knott and Riggin (41), Hartmann (42) for ICR drift cells and by Sharp *et al.* (43), McIver *et al.* (44), Comisarow (45) and Dunbar (46) for ion traps.

An example of the complexity of the frequency variations in the cubic cell is given in the extreme by the splitting of a peak into a doublet for ions excited to very large orbits. A similar phenomenon was noted by Marshall (47). In the example reported here, the ions were excited by a 0.385 volt peak-to-base RF burst in a 0.0254 m cubic cell with a 1 volt trap and a magnetic field of 1.2 T. Two maxima are discernible for ion-cyclotron-orbit sizes larger than the "optimal" orbit size at about 760 μsec excitation time. Local centroids are measurable; one increases by ca. 50 Hz and the other

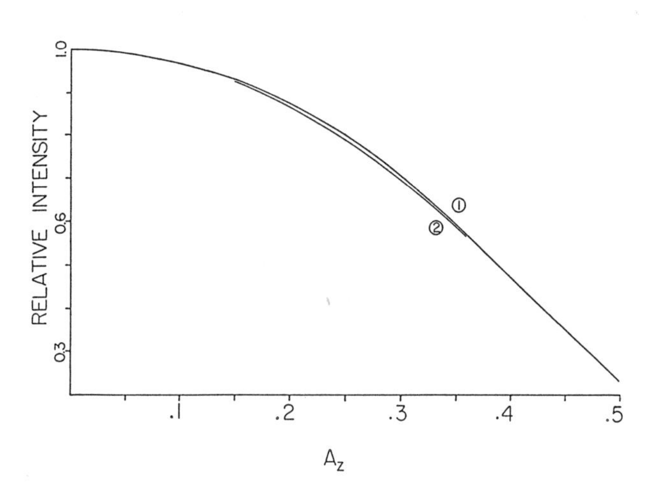


Figure 6. Relative fundamental harmonic amplitude as a function of z-oscillation amplitude. (1) $A_r$ = 0.25 when $A_z$ = 0. (2) $A_r$ = 0.05 when $A_z$ = 0. (Reproduced with permission from Ref. 33. Copyright 1986 Elsevier Science Publishers B.V.)

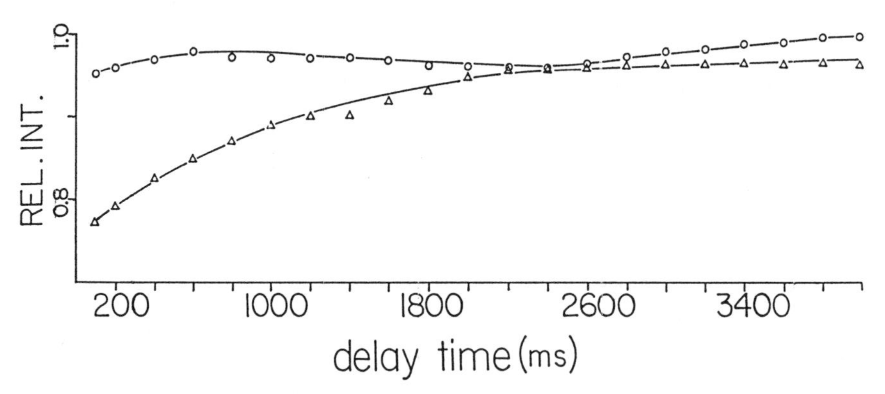


Figure 7. The relative ion signal intensity of an initial 100,000 benzene ions vs. time under conditions of no charge exchange at 1.0 V trap (O) with and (Δ) without employment of ion compression. The initial trap voltage was 0.125 V for (O). (Reproduced with permission from Ref. 33. Copyright 1986 Elsevier Science Publishers B.V.)

decreases by more than 180 Hz away from the "optimal" orbit size frequency as the excitation time is increased.

If the magnetron mode is negligible in size, then the cyclotron frequency shift caused by the radial electric fields is approximately proportional to the magnitude of the radial electric field component divided by the radius, $-E_r/r$ (48). Based on Dunbar's treatment (46), which shows a decrease in the average $E_r/r$ in the z = 0 plane, one would expect ions with low z-mode amplitudes to increase in frequency as their orbit sizes are increased. Evaluation of radial electric trapping fields at larger z and radial positions shows a large increase suggesting that the frequencies of ions with high z-mode amplitudes would decrease with increasing orbit size. Thus, we attribute the high-frequency peak to ions with low z-mode amplitudes and the low-frequency peak to ions with high z-mode amplitudes.

Because low amplitude RF burst waveforms do not significantly modify the z-mode amplitudes of ions, the intensities would be expected to reflect the z-mode amplitude distribution just before excitation. This gives us one means of checking the above hypothesis: by allowing the z-mode amplitudes to relax via ion-molecule collisions, the relative peak intensities should change. Indeed, at long delay, the high frequency peak increases at the expense of the low frequency peak.

The observations that there is an "optimum" orbit size and that peaks split for orbits not too much larger than the optimum orbit suggest that the optimum orbit occurs because of special circumstances. One possible circumstance is a coincidence of frequencies for ions with low and high z-mode amplitudes so that if there are mass discriminating differences in the way the ions populate the trap or in the way ions are excited, then systematic mass measurement errors can be expected. Excitation of the cyclotron mode does produce a spread in cyclotron radii, and mass discriminating z-mode excitation is discussed elsewhere in this chapter. Thus, frequency variations that cause systematic mass errors are due in part to trap field inhomogeneities. These effects are evident at low ion populations and may be due in part to excitation induced ion cloud deformation which increases with ion number.

Systematic errors that increase with ion number occur, perhaps, because assumptions made in the model are being violated in practice. One of the basic assumptions of the space charge theory of Jefferies (49), from which the calibration law was derived, is that a thermal ion cloud in a cubic Penning trap takes the shape of an oblate ellipsoid with a uniform distribution of charge. If so, the space charge potential should be quadratic, thus producing a constant contribution to $E_r/r$. If excitation were to alter the shape of the ion cloud, and consequently the linearity of the space charge electric field, then the linear model no longer applies, and systematic errors may occur. Altering the shape of the ion cloud is not unexpected. The orbits of ions are known to be large after excitation compared to before excitation, and calculations show that the cyclotron orbits vary by as much as 50% of the largest orbit for RF burst (50) and chirp (25) excitations. z-Excitation also affects the ion distribution in the cell as is discussed earlier in this chapter.

3. Tools: To determine the accuracy of a derived calibration law, computed masses of ions are compared to known values by using the calibration shown in equation (2) for the cubic cell or that shown in equation (3) for the hyperbolic.

Hyperbolic cell $$f^2 = a/m^2 + b/m \qquad (3)$$

The values of a and b were determined empirically by measuring the frequencies of six reference masses and fitting mass-frequency data to the form of the equation by means of a least-squares procedure. By using empirical values of a and b and the measured ion frequencies, the masses of the six ions were calculated. A statistical evaluation between the known masses and the experimentally determined masses was then made to determine the accuracy of the calibration law. Systematic errors were observed as is described in (40).

4. Possible Solutions: New cell designs have been investigated as possible solutions for improving accurate mass measurements. Efforts have been directed at minimizing frequency complexity by linearizing the trapping fields and reducing space charge effects by spreading out the ions. The reduction of space charge effects would also permit improvement in dynamic range.

One possibility is the rectangular cell proposed by McIver *et al.* (29). The cell is elongated along the z-axis so that the ion frequencies experience less perturbation associated with the trapping fields. Also because of the large volume in the cell, more ions can be stored with less ion-ion coupling effects than in the cubic cell.

A recent effort of this laboratory has resulted in the introduction of a hyperbolic Penning trap (31). The cell consists of two end caps and one ring electrode similar to the design of Byrne and Farago (51) (see Figure 8).

Ions are formed by an electron beam offset at 1/2 $r_o$ because ions cannot be excited at the center of the cell. This is done also to allow the largest possible cyclotron orbit between the z-axis and the ring electrode. The cyclotron mode of the ions is excited across the ring and end caps, and image currents are detected by using a balanced bridge circuit (52).

The hyperbolic cell offers several advantages for accurate mass determinations. 1) The frequencies of fundamental modes of the ion motion are independent of the location of the ion in the cell at least for the low ion limit. This is important because frequencies are known to be influenced by the location of the ions in cubic cells. 2) The excitation electric field has a simple form identical to that of the trapping field. Thus, the operation of the trap is the same everywhere in the interior permitting most of the volume to be used for storage of larger ion populations with reduced space charge effects. 3) Harmonic effects (31,13b) that complicate spectra are removed.

The mass calibration law gives mass measurement accuracy of ca. 2 ppm and a precision of ca. 1 ppm. Errors are still systematic, but the measurements are much less sensitive to space charge effects than those made with the cubic cell. One possible cause of the systematic errors may be magnetic field inhomogeneity caused by

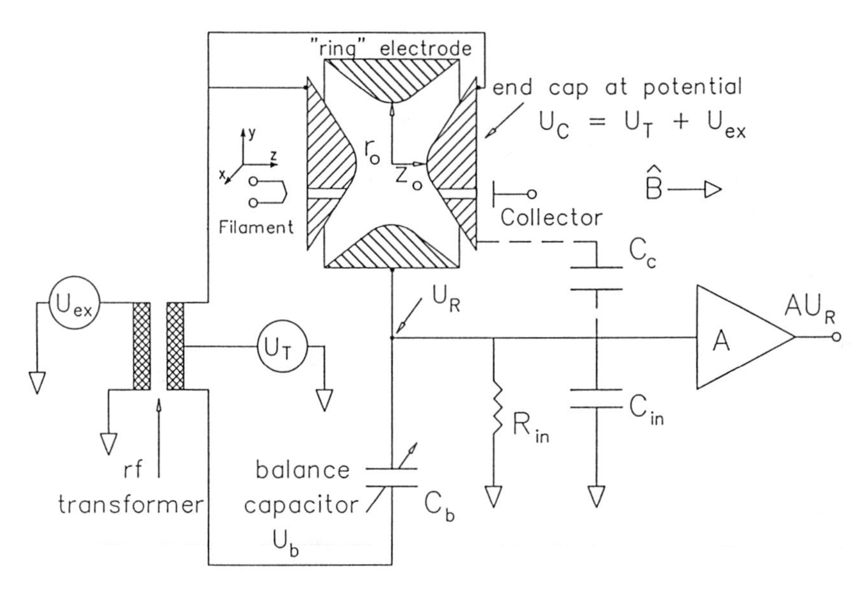


Figure 8. Hyperbolic Penning cell imbedded in a balanced bridge network.

local magnetic fields induced by machining the cell from stainless steel.

Another cell design may incorporate the use of segmented trap plates in the cubic cell (Figure 4) (53). Segmenting the trap plates is expected to minimize trapping field inhomogeneities.

Poor Resolution for High Mass Ions. 1. Problem Definition and Model: The magnitude of the electric forces, both trapping and coulombic, relative to the magnetic force increases with increasing mass. As a result one would expect that any motion problems resulting from electric fields would also increase with ion mass. This may be one of the causes of the low resolution detection of high mass ions (54), particularly organic ions, by FTMS. This problem definition is not intended to detract from the remarkable accomplishments of Hunt, Russell and coworkers which are reported elsewhere in this volume.

The study of high mass ion dynamics is complicated by the relatively poor control over ion populations by methods used to generate high-mass ions (e.g., laser desorption, SIMS) and by the possibility that chemical behavior of the ion (e.g., slow decompositions) will interfere with efforts to gain information about dynamical behavior. We are proposing a scaling technique that may be used to gain experimental insight into the dynamics of high-mass ions. Scaling (55) involves extrapolating information about the dynamics of high-mass ions at high magnetic fields by studying experimentally low mass ions at lower magnetic fields.

2. Model: Consider the phenomenological equation (56) of the motion of the ion to be studied:

$$\frac{d^2\vec{x}}{dt^2} = \frac{q}{m}\,\vec{E}(\vec{x}) + \frac{d\vec{x}}{dt} \times \vec{B} - \zeta\frac{d\vec{x}}{dt} \quad (4)$$

By introducing the coordinate transformations

$$t = (q/m)^{1/2}t \quad (5)$$

$$\vec{\chi}(t) = \vec{x}(t) \quad (6)$$

and parameter changes

$$\vec{\beta} = \vec{B}/(m/q)^{/2} \quad (7)$$

$$\nu = (m/q)^{1/2}\zeta \quad (8)$$

the phenomenological equation becomes

$$\frac{d^2\vec{\chi}}{d\tau^2} = \vec{E}(\vec{\chi}) + \frac{d\vec{\chi}}{d\tau} \times \vec{\beta} - \nu\frac{d\vec{\chi}}{d\tau} \quad (9)$$

Note that the electric field as a function of ion position remains unchanged. The solution, $\vec{\chi}$, of the scaled equation represents the physical solution, $\vec{x}$, for any situations where $\vec{B}/(m/q)^{1/2} = \vec{\beta}$ and $\zeta(m/q)^{1/2} = \nu$. The $\vec{x}$ is recovered from $\vec{\chi}$ by scaling the time variable. Thus, it can be seen that the behavior of two situations for which

$$\vec{B}/(m_2/q)^{1/2} = \vec{B}/(m_2/q)^{1/2} \quad (10)$$

must be the same after the appropriate time scale adjustment is made. This would suggest high mass behavior at high magnetic fields can be studied with low mass ions at low magnetic fields.

Tool: As a tool, scaling requires that experimental conditions be chosen that imitate the variable changes required to scale the equations for a high mass ion at a high magnetic field into equations for a low mass ion at a lower magnetic field.

As an example of how mass scaling would be used as a tool, consider that an investigation directed nominally at the dynamics of an ion of m/z 2000 at 3 T can be made by using the benzene molecular ion at 0.59 T. By analogy, it should be possible to study the behavior of low mass ions and relate that information directly to the behavior of high mass ions at high magnetic fields.

The scaling approach is at the proposal stage, and we have no results at this time and no detailed solutions to suggest for the problem.

Low Mass Resolution Ion Selection for MS-I in MS/MS Experiments. 1. Problem Definition: The advantages of FTMS for MS/MS and particularly for $(MS)^n$ are often touted, yet few reports of problem-solving in chemical analysis have been made. The mass resolution for daughter ions can be superb in the MS/MS mode, as has been demonstrated by Wilkins et al. (57) and Freiser et al. (58). However, the resolution of MS-I (for parent ion selection) in cubic cells is poor. Moreover, the energy for collisional activation is limited to the cyclotron mode energy and this limits applications involving high mass ions. Finally, the original population of ions is ejected, except for the ion of interest. This latter disadvantage pertains also to sector and quadrupole tandem instruments, but it need not apply to FTMS. That is, it should be possible to retain some of the "multichannel advantage" of FTMS in both the MS-I and the MS-II modes by not throwing away most of the ions in the first MS step. It may be highly desirable to obtain information on other ionic species from the original ion population particularly when sample sizes are limited in quantity or when ion production is poor and a limiting factor.

2. Model: The difference between the mass resolution of daughter ions in the second MS step and the mass resolution with which one can select the parent ion in the first MS step is due to a difference between the operations of these two steps. In the second MS step, the signals that result in the mass spectrum are observed after the conclusion of the excitation. So the width of a particular peak in the spectrum is determined by trap field inhomogeneities of the regions through which the ions pass after the excitation is complete. Because only collisional processes will change the amplitudes of the z, magnetron, and cyclotron modes during this time, one would expect the ion trajectories to be contained in a proper subset of the volume of the cell. In the first MS step, the ejection resolution describes the limits for accelerating one ion species to the point of ejection while leaving a neighbor mass species essentially undisturbed at the conclusion of the excitation. The act of accelerating an ion that is initially bounded close to the z-axis to an ion with a large cyclotron mode

for ejection requires the ion to move through a large portion of the trap volume. As a result, the motion of the ion is affected by the full extent of the trap field inhomogeneities. By this reasoning, the resolutions of the two MS steps would be different, and the ejection resolution of the first MS step would be poorer.

For low ion populations, a first estimate of achievable ejection resolution might be obtained from the cyclotron frequency spread that occurs over the range of cyclotron orbit radii through which the ion must pass to be ejected. This is based on the notion that an ejection waveform that is just adequate to eject one ion must have a frequency spectral peak that is at least as wide as the above spread of frequencies. Such a waveform would then excite, at least to some extent, all ions with frequencies falling within the width of the peak, thus limiting the ejection resolution. For ions with low z-mode amplitudes, we can use Dunbar's (46) approximate expression for the average radial field strength,

$$E_r \simeq 2\alpha(V_T/a^2)\ r\ [1-2.8(r^3/a^3)] \tag{11}$$

to obtain the frequency spread between the smallest and largest orbit:

$$\Delta\omega_{eff} \simeq 0.7(qB/m_c)(r_m/a)^3 \tag{12}$$

Here, B is the magnetic field strength, $m_c$ is the critical mass (40), q is the charge of the ion in question, $r_m$ is the maximum radius required for ejection, a is the inside length of the cubic cell, r is a radial displacement, $V_T$ is the voltage applied to the trap plates, and $\alpha$ = 1.3871 (49) a geometry constant.

In the derivation of the approximate expression for $\Delta\omega_{eff}$, use was made of

$$\Delta_{eff} = 1/2\ qB/m\ \{1 + [1-4(m/q)(E_r/r)/B^2]^{1/2}\} \tag{13}$$

for the cyclotron frequency of an ion of mass, m, with an orbit centered on the z-axis, and

$$m_c/q = q^2B^2/8\alpha\nu_T \tag{14}$$

for the critical mass (upper bound on the range of ion masses for which the ion trajectory is stable in the trap).

From the expression for $\Delta\omega_{eff}$, the ejection resolution becomes

$$m/\Delta m \simeq \omega_{eff}/\Delta\omega_{eff} \simeq 1.429(m_c/m)(a/r_m)^3 \tag{15}$$

As an example, in a 2.54 cm cubic trap with B = 1.2 T and $V_T$ = 1 V, the above estimate for the ejection resolution gives 1.59 K.

3. Tool: RF burst waveforms serve as a convenient means for determining the ejection resolution limits that are due to effects intrinsic in the operation of the cubic trap. More general computed waveforms (59) are obtained via inverse discrete Fourier transformation. These computed waveforms are just a linear superposition of a finite number of RF bursts. As a result, it is proposed that the best performance obtained with RF bursts anticipate the best performance obtained with computed waveforms,

when these waveforms are used for ejection. In particular, the rf burst is the special case in the class of all computed waveforms that is best suited for ejection of a single mass.

The MS-1 resolution is determined through the use of ejection spectra. These spectra are produced in an experiment using two rf bursts, the first is the test ejection waveform and the second quantitates the number of unejected ions. The spectra are plots of ion signal intensity versus the frequency of the test waveform as the frequency is stepped through the resonance of the ion. At the ion resonance, the plot shows a dip having a width that indicates the ejection resolution of the test waveform.

The first step in the measurement of the intrinsic ejection resolution limits of the trap is to find the minimum ejection waveform time - amplitude product for which the dip minimum shows complete loss of ion signal. This is done by varying the amplitude of the test waveform and fixing the time at ~1 msec to assure the peak width in the frequency spectrum of the test waveform is wide enough to cover the range of cyclotron frequencies of the ejected ion.

Then, in steps, the test waveform time is increased while maintaining the original time-amplitude product until the dip minimum of the ejection spectrum acquired for each step no longer shows a complete loss of ion signal. At this step, the half-height width of the dip is used to determine the resolution. If the trap were functioning as an ideal trap, that is, as a linear device with no collisional damping, the test waveform time could be indefinitely increased without ever showing incomplete loss of ion signal at the dip minimum.

By using this technique, the ejection resolution measured for the example given in the "Model" section is certainly no better than 2.4 K. This compares with a normal mass resolution of 30 K under the same trap conditions.

4. Proposed Solution: We suggested earlier (9) the possibility of an intermediate cell in MS/MS applications for storing all ions from which the one of interest is selected. This may be accomplished by assembling two adjoining cubic cells in a manner similar to that in (60) except the partitioning or center trap plate would not contain an orifice but rather be a solid disk (see Figure 9). The disk would be centered on the z-axis and have a diameter of 1/2 the cell dimension. The disk will be held in place by a highly transparent metallic screen or grid that is insulated from the disk so that it can be operated at a potential different than that of the disk.

Electrons (or any ionizing beam) would be admitted from the left through an orifice in the trapping plate opposite the disk. They would strike the disk after traversing the first cell of the tandem. Ionization of a pulsed sample (from a pulsed valve) would begin the experiment although ions may be introduced from other traps by using the partitioning principle of the Nicolet dual cell (60) or even more effectively by employing ion transfer techniques similar to those described over twenty years ago (51).

Once the ion population is formed and stable and the neutral population is again low, the ions to be activated are accelerated to a cyclotron orbit slightly larger than the disk. Up to this point, the supporting screen and disk have been at the same trapping

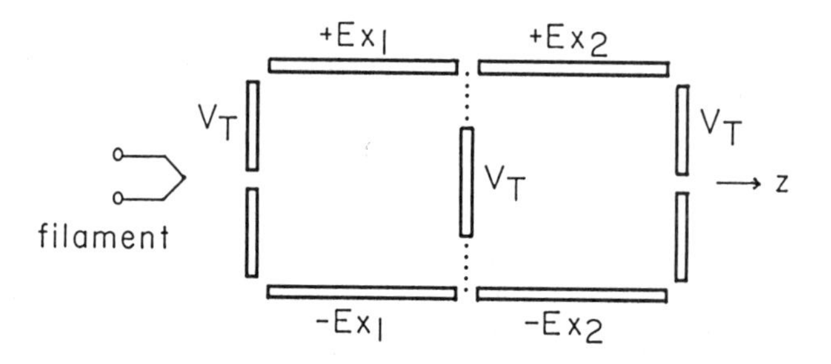


Figure 9. Proposed segmented cell with disk shaped center trap plate.

potential, but now the screen is pulsed in a fashion similar to that described in (60,61) to transfer the excited ion to an adjoining cell. The z-mode motion of the unexcited ions is not disturbed sufficiently to climb the potential barrier still provided by the disk.

The selected ions now in the adjoining cell are returned nominally to the z-axis by using the method of Marshall (62). Even though this deexcitation may not be perfect, it should be sufficiently adequate that the next step of ion activation can be performed.

Ion activation takes place by admitting particles (photons, ions, electrons) into the cell containing the ion of interest. These activating particles are injected along the z-axis so that their trajectory is interrupted by the disk, and the ions in the first cell are preserved.

The segmented tandem cell should have the following advantages: 1) The original population of ions is not lost by ejection as is currently the case. 2) The selected ions are activated along the z-axis, and this should lead to better performance in the MS-II process. 3) The ion of interest can be selected at higher mass resolution because the selection radius is kept small (increased $a/r_m$ in Equation 15) and because better trapping fields can be generated by the disk/screen arrangement (the outer segment trap plate potential would be set for minimum ion frequency spread during excitation). 4) The activation energy is not limited to the cyclotron mode energy. 5) The original population of ions is shielded from the activation process.

## Conclusions

Most FTMS instrument and method development research has been focussed on demonstration experiments. Examples include coupling FTMS with various sample introduction schemes (e.g., GC, LC, supercritical fluid chromatography), sample ionization (e.g., LD, pulsed SIMS, Cf-252 PDMS, etc.), and demonstrating application to various interesting classes of chemical compounds. These demonstrations are useful because they are indications of the potential of the technique. However, few reports of the routine use of FTMS for trace analysis, for accurate mass, and for structure determination of unknowns have yet appeared. One reason is that FT mass spectrometers are not widely spread in the hands of users. Another is that FTMS is not yet routine. Most of the demonstration experiments have been done in expert laboratories by committed and highly focussed graduate students and postdoctoral researchers.

We propose that an essential route to the routine utilization of FTMS is to focus on fundamental issues that face any spectrometric method; that is, issues dealing with accurate determination of peak position, peak shape, and peak height. In this paper, five problems were introduced. With regard to peak height measurements are problems of nonuniform tuning owing to inadvertant z-ejection of ions and temporal variation of ion signals because of ion cloud relaxation phenomena. Systematic mass measurement errors are an example of problems in determining peak positions. Less than optimum resolutions for high mass ions and for ejecting ions in MS/MS-type experiments are examples of peak-shape difficulties. In all cases, approaches to solving the problems are given and, in many cases, actual solutions are presented.

Acknowledgments

This work was supported in part by the Midwest Center for Mass Spectrometry (NSF Grant CHE-8620177).

## Literature Cited

1. Svec, H.J. Int. J. Mass Spectrom. Ion. Proc., 1985, 66, 3.
2. Thomson, J.J., Rays of Positive Electricity and the Application to Chemical Analyses; Longmans Green: London, 1913.
3. Dempster, A.J., Phys Rev, 1918, 11, 316.
4. Aston, F.W., Philos. Mag. VI, 1919, 18, 707.
5. Bartky, W.; Dempster, A.J., Phys. Rev., 1929, 33, 1019.
6. Cottrell, J.S.; Greathead, R.J., Mass Spectrom. Rev., 1986, 5, 215.
7. a. Bateman, R.H.; Burns, P.; Owen, R.; Parr, V.C., Adv. Mass Spectrom., 1986, 1013, 863. b. Bateman, R.H.; Owen, R.; Parr, V.C.; Wood, D., Proceedings of the 34th Annual Conference on Mass Spectrometry and Allied Topics, 232, 1986.
8. Comisarow, M.B.; Marshall, A.G., Chem. Phys. Lett., 1974, 25, 282.
9. a. Comisarow, M.B.; Marshall, A.G., J. Chem. Phys., 1976, 64, 110. b. Gross, M.L.; Rempel, D.L., Science, 1984, 226, 261.
10. Ogata, K.; Matsuda, H., Z. Naturforsch. Teil A, 1955, 10, 843.
11. Allemann, M.; Kellerhals, Hp.; Wanczek, K.-P., Int. J. Mass Spectrom. Ion Phys., 1983, 46, 139.
12. Laude, D.A.,Jr.; Johlman, C.L.; Brown, R.S.; Weil, D.A.; Wilkins, C.L., Mass Spectrom. Revs., 1986, 5, 107.
13. a. Comisarow, M.B., J. Chem. Phys., 1978, 69, 4097. b. Nikolaev, E.N.; Gorshkov, M.V., Int. J. Mass Spectrom. Ion Proc., 1985, 64, 115.
14. a. Amster, I.J.; McLafferty, F.W.; Castro, M.E.; Russell, D.H.; Cody, R.B., Jr.; Ghaderi, S., Anal. Chem., 1986, 58, 483. b. Russell, D.H., Mass Spectrom. Rev., 1986, 5, 167.
15. Aberth, W., Anal. Chem., 1986, 58, 1221.
16. McCrery, D.A.; Sack, T.M.; Gross, M.L., Spectroscopy: An International Journal, 1984, 3, 57.
17. Ghaderi, S.; Kulkarni, P.S.; Ledford, E.B., Jr.; Wilkins, C.L.; Gross, M.L., Anal. Chem., 1981, 53, 428.
18. Laude, D.A., Jr.; Johlman, C.L.; Brown, R.S.; IJames, C.F.; Wilkins, C.L., Anal. Chim. Acta, 1985, 178, 67.
19. Sack, T.M.; Gross, M.L., Anal. Chem., 1983, 55, 2419.
20. Sack, T.M.; McCrery, D.A.; Gross, M.L., Anal. Chem., 1985, 57, 1290.
21. McCrery, D.A.; Peake, D.A.; Gross, M.L., Anal. Chem., 1985, 57, 1181.
22. Peake, D.A., Ph.D. Thesis, University of Nebraska - Lincoln, 1986.
23. Huang, S.K.; Rempel, D.L.; Gross, M.L., Int. J. Mass Spectrom. Ion Proc., 1986, 72, 15.

24. Sack, T.M., Ph.D. Thesis, University of Nebraska-Lincoln, 1985.
25. Huang, S.K.; Rempel, D.L.; Gross, M.L., unpublished work.
26. Beauchamp, J.L.; Armstrong, J.T., Rev. Sci. Instrum., 1969, 40, 123.
27. Sherman, M.G.; Francl, T.J.; McIver, R.T., Jr., Proceedings of the 31st Annual Conference on Mass Spectrometry and Allied Topics, 1983, 400.
28. Wang, T.C.L.; Marshall, A.G., Int. J. Mass Spectrom. Ion Proc., 1986, 68, 287.
29. Hunter, R.L.; Sherman, M.G.; McIver, R.T., Int. J. Mass Spectrom. Ion Phys., 1983, 50, 259.
30. Lee, S.H.; Wanczek, K.-P.; Hartman, H., Advances in Mass Spectrometry, 1980, 8B, 1645.
31. Rempel, D.L.; Ledford, E.B.,Jr.; Huang, S.K.; Gross, M.L., Anal. Chem, In press.
32. Clow, R.P.; Futtrell, J.H., Int. J. Mass Spectrom. Ion Phys., 1972, 8, 119.
33. Rempel, D.L.; Huang, S.K.; Gross, M.L., Int. J. Mass Spectrom. Ion Proc., 1986, 70, 163.
34. (a) Riggin, M.; Woods, I.B., Can J. Phys., 1974, 52, 456. (b) Bloom, M.; Riggin M., Can. J. Phys., 1974, 52, 436.
35. (a) Dehmelt, H.G., Adv. At. Mol. Phys., 1967, 40, 53; 1969, 5, 109. (b) Church, D.A.; Dehmelt, H.G., J. Appl. Phys., 1969, 40, 3421. (c) Church, D.A.; Mokri, B., Z. Phys., 1971, 224, 6. (d) Heppner, R.A.; Walls, F.L.; Armstrong, W.T.; Dunn, G.H., Phys. Rev. A, 1976, 13, 1000.
36. Dunbar, R.C., Int. J. Mass Spectrom. Ion Proc., 1984, 56, 1.
37. Ledford, E.B.,Jr.; Ghaderi, S.; White, R.L.; Spencer, R.B.; Kulkarni, P.S.; Wilkins, C.L.; Gross, M.L., Anal. Chem., 1980, 52, 463.
38. Allemann, M.; Kellerhals, Hp.; Wanczek, K.-P., Chem. Phys. Lett., 1981, 84, 547.
39. Francl, T.; Sherman, M.G.; Hunter, R.L.; Locke, M.J.; Bowers, W.D.; McIver, R.T., Jr., Int. J. Mass Spectrom. Ion Proc., 1983, 54, 189.
40. Ledford, E.B.,Jr.; Rempel, D.L.; Gross, M.L., Anal. Chem, 1984, 56, 2744.
41. Knott, T.F.; Riggin, M., Can. J. Phys., 1974, 52, 426.
42. Hartmann, H.; Chung, K.-M.; Baykut, G.; Wanczek, K.-P., J. Chem. Phys., 1983, 78, 424.
43. Sharp, T.E.; Eyler, J.R.; Lie, E., Int. J. Mass Spectrom. Ion Phys., 1972, 9, 421.
44. a. Sherman, M.G.; Francl, T.J.; McIver, R.T., Jr., Proceedings of the 31st Annual Conference on Mass spectrometry and Allied Topics, 1983, 400. b. Francl, T.J.; Sherman, M.G.; Hunter, R.L.; Locke, M.J.; Bowers, W.D.; McIver, R.T., Jr., Int. J. Mass Spectrom. and Ion Proc., 1983, 54, 189-199.
45. Comisarow, M.B., in Lecture Notes in Chemistry, 31, ICR Spectrometry II, H. Hartmann and K.-P. Wanczek, eds., Springer Verlag, Berlin, 1982, 484.
46. Dunbar, R.D.; Chen, J.H.; Hays, J.D., Int. J. Mass Spectrom. Ion Proc., 1984, 57, 39.
47. Marshall, A.G., J. Chem. Phys., 1971, 55, 1343.

48. Davidson, R.C., Theory of Non Neutral Plasmas, W.A. Benjamin, Reading, Massachusetts, 1974.
49. Jeffries, J.B.; Barlow, S.E.; Dunn, G.H., Int. J. Mass Spectrom. Ion Proc., 1983, 54, 169.
50. Rempel, D.L.; Huang, S.K.; Gross, M.L., Int. J. Mass Spectrom. Ion Proc., 1986, 70, 163.
51. Byrne, J.; Farago, P.S., Proc. Phys. Soc., 1965, 86, 801.
52. McIver, R.T., Jr.; Hunter, R.L.; Ledford, E.B., Jr.; Locke, M.J.; Francl, T.J., Int. J. Mass Spectrom. Ion Phys., 1981, 39, 65.
53. Huang, S.K.; Rempel, D.L., Proceedings of the 34th Annual Conference on Mass Spectrometry and Allied Topics, 1986, 167.
54. a. Amster, I.J.; McLafferty, F.W.; Castro, M.E.; Russell, D.H.; Cody, R.B., Jr.;Ghaderi, S., Anal. Chem., 1986, 58, 483.
b. Hunt, D.F., private communication.
55. Arnold, V.I., Geometric Methods in the Theory of Ordinary Differential Equations, Springer Verlag, 1983.
56. Beauchamp, J.L., J. Chem. Phys., 1966, 46, 1231.
57. White, R.L.; Wilkins, C.L., Anal. Chem., 1982, 54, 2211.
58. Cody, R.B.; Freiser, B.S., Anal. Chem., 1982, 54, 1431.
59. Marshall, A.G.; Wang, T.C.L.; Ricca, T.L., J. Am. Chem. Soc., 1985, 107, 7893.
60. Cody, R.B.; Kinsinger, J.A.; Ghaderi, S.; Amster, J.J.; McLafferty, F.W.; Brown, C.E., Anal. Chim. Acta., 1985, 178, 43.
61. Schwinberg, P.B.; Van Dyck, R.S., Jr.; Dehmelt, H.G., Phys. Rev. Lett., 1981, 47, 1679.
62. Marshall, A.G.; Wang. T.-C.; Ricca, T.L., Chem. Phys. Lett., 1984, 105, 234.

RECEIVED June 15, 1987

# Chapter 4

# Application of the Dual-Cell Fourier Transform Mass Spectrometer

**Robert B. Cody, Jr., and James A. Kinsinger**

**Nicolet Analytical Instruments, 5225 Verona Road, Madison, WI 53711**

The differentially-pumped, dual cell geometry has improved the performance of Fourier transform mass spectrometry (FTMS) for a variety of analyses, including combined gas chromatography/mass spectrometry, laser desorption mass spectrometry, and MS/MS. The ability to accommodate higher source pressures has also permitted the coupling of supercritical fluid chromatography with FTMS. The dual cell geometry has demonstrated improved resolution and accurate mass measurements under analytical conditions; added benefit results from isolating sample ions from reactive neutrals.

The potential benefits of Fourier transform mass spectrometry have been evident ever since the introduction of the technique by Comisarow and Marshall in 1974 [1], but it is only in recent years that major developments in the technology have caused an increased acceptance of FTMS into the mainstream of analytical mass spectrometry. In particular, the development of differentially-pumped systems including the tandem quadrupole-FTMS [2], the dual-cell geometry [3,4], and external sources [5], have overcome the difficulties associated with the requirement that FTMS analysis be accomplished at low pressures (e.g., less than $10^{-7}$ torr) for best results. Efforts to develop FTMS for high mass detection [6,7], in particular, the work by Shabanowitz, et al. [8] with the detection of insulin and cytochrome C, have demonstrated that large ions can be formed with sufficiently long lifetimes to be detectable on the FTMS timescale, proving that FTMS is a viable technique for the analysis of molecules in the "middle-mass" range (e.g., 1,000-15,000 amu). The tailored-excitation method developed by Marshall [9,10] provides a means to overcome several limitations of current ion excitation methods, and promises to bring about significant improvements in the overall performance of FTMS, particularly FTMS/MS [11].

In this article, we describe some of the recent developments and applications of the dual-cell Fourier transform mass

0097–6156/87/0359–0059$06.50/0

spectrometer to a variety of chemical problems and mass spectrometric techniques.

## Experimental

All experiments were performed using a Nicolet Analytical Instruments FTMS-2000 dual-cell Fourier transform mass spectrometer with optional GC and laser desorption interfaces. The FTMS-2000 dual cell is specially constructed of stainless steel with low magnetic susceptibility. This permits very efficient ion transfer between the source and analyzer cells, if the cells are properly aligned in the magnetic field.

The superconducting magnet was operated at a field strength of 3 Tesla. This instrument has been described in previous publications [4] to which the reader is referred for a more detailed description. Specific details of the various experiments described in this report will be included in the individual sections pertaining to those experiments.

## Results and Discussion

Gas Chromatography/FTMS. The differentially pumped, dual-cell geometry has permitted the practical realization of combined gas chromatography/Fourier transform mass spectrometry (GC/FTMS) by providing a means to deal with the higher pressures associated with the GC carrier gas [12]. Using the dual-cell instrument, we have shown that it is possible to obtain high resolution mass spectra, even for small sample quantities. For example, we were able to detect ions from a 125 pg splitless injection of naphthalene at a resolution of 15,000 over the mass range 120 to 130 amu with a signal-to-noise ratio of 10:1[4].

Once the carrier gas pressure has been eliminated as a factor, the major factor affecting the resolution obtainable for GC/FTMS spectra is the trade-off between the mass spectral data acquisition time (observation time and number of signal-averaged transients) and the data acquisition rate required for high resolution capillary gas chromatography. In order to maximize mass spectral resolution, it is desirable to observe the transient signal for as long a time as possible. However, the width of chromatographic peaks places a limit on how long an observation time may be employed without affecting chromatographic resolution. Typically, peaks in capillary gas chromatography are about 3 seconds wide, which means that, ideally, mass spectra should be acquired about once per second.

In order to demonstrate that it is possible to obtain very high mass resolution without sacrificing chromatographic resolution, we made a splitless injection of an activity mix (SGE Activity Mix A) containing approximately 50 ng each of naphthalene, 2,6-dimethylphenol, and 2,4-dimethylaniline. The mass spectrometer was set up to operate in heterodyne (high resolution) mode [13] to monitor masses within the range 119-129 amu. Two transients were collected and signal-averaged per spectrum, with 64K data points acquired per data set.

Figure 1 shows the molecular ion of naphthalene, detected at a resolution of 340,000. The total acquisition time per transient

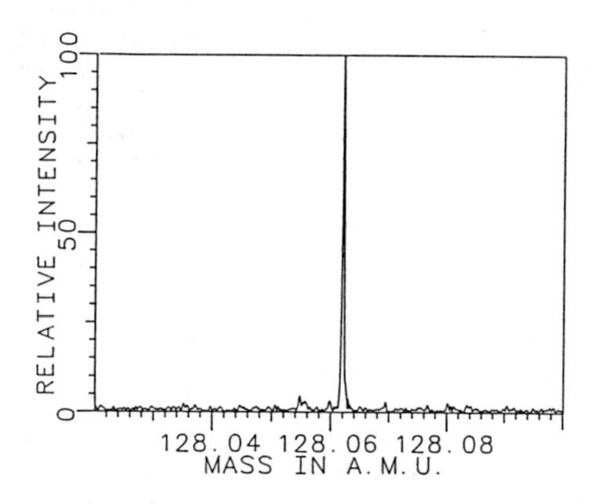


Figure 1. GC/FTMS: Naphthalene molecular ion detected at a resolution of 342,000.

was 1.2 seconds. By observing the transients, it could be seen that only one of the two averaged transients contributed signal. This suggests that the optimum conditions would be to collect only one transient per spectrum, which would reduce the total acquisition time to 1.2 seconds per spectrum. This data shows that it is possible to obtain ultra-high mass resolution (greater than 100,000) on a time scale that is consistent with high chromatographic resolution.

The major application of high resolution mass spectrometry is for obtaining accurate mass measurements in order to determine elemental compositions. The present FTMS calibration equation, derived by Ledford, et al. [14] shows that the mass-to-charge ratio m/z of a given ion is related to the magnetic field, B, the electric field, E, and the measured frequency, F, by a relation having the form:

$$m/z = k_1 + B/f + k_2 E/f^2$$

where $k_1$ and $k_2$ are constants, and the electric field term is related to the trap voltage and the number of ions in the cell. Wilkins, et al. have demonstrated accurate mass measurements in the low part-per-million range for GC/FTMS, using a calibration performed just prior to the sample analysis [15]. The success of this method is based upon the fact that the calibration is performed under conditions which are virtually identical to those employed for the sample measurement, and hence the total number of ions is approximately the same for both measurements.

It is also possible to improve the accuracy of the mass calibration by making use of a peak of known composition to correct the electric field term for changes in the number of ions [16]. In other words, if the identity of some component of the mixture is known by its retention time and/or mass spectrum, a peak of known mass in the mass spectrum may be used to apply a correction factor, x, to the electric field term in the calibration equation, which may be rewritten as:

$$m/z = k_1 B/f + k_2 (E+x)/f^2$$

where m/z is chosen to be the mass of the known peak, and f is the measured frequency of that peak. Subsequent mass measurements using the corrected calibration equation will have a higher mass measurement accuracy than those using the uncorrected equation.

This is illustrated for the activity mix in Table 1. Masses for ions within the 119-129 amu mass range were measured for the three compounds, using a calibration taken prior to the measurement. The average deviation for six ions is found to be 6.9 ppm. If we were certain of the identity of the molecular ion from naphthalene (e.g., if naphthalene had been deliberately added to the mixture), we could use the molecular ion as an "internal standard" to determine the appropriate correction to the electric field term in the calibration. From Table 1, we see that the result of this "one point" correction is a significant improvement in the mass measurement accuracy, with the average deviation in the measurements of the remaining five ions being only 0.8 ppm

This sort of correction should be helpful for improving specific mass measurements for GC/FTMS analyses where it is possible to add a calibrant compound to the mixture, or where some component is known to be present. Unlike the method where a static

**Table I.**
**Mass Measurement Accuracy from High Resolution GC-FTMS Run**

| COMPOUND | PRIOR CALIBRATION | | ONE POINT CALIBRATION | |
|---|---|---|---|---|
| | Mass (amu) | Deviation (ppm) | Mass (amu) | Deviation (ppm) |
| 2,6-Dimethylphenol | 122.072316 | 7.0 | 122.072674 | -0.5 |
| 2,4-Dimethylaniline | 120.080473 | 7.1 | 120.080819 | -0.4 |
| | 121.088187 | 8.0 | 121.088539 | 0.5 |
| Naphthalene | 126.046190 | 6.0 | 126.046572 | -1.4 |
| | 127.053994 | 6.1 | 127.054382 | -1.2 |
| | 128.061660 | 7.3 | 128.062054 | -0.01 |
| | **AVERAGE** | **6.9** | | **0.7** |

calibration compound is present, the "internal calibrant" here is separated by the gas chromatograph from the sample compounds, and does not interfere with the mass spectra of the remaining compounds. It should be noted that the best results are achieved when the ion chosen to determine the correction is close in mass to the ions to be measured.

The additional information available from high resolution GC-MS may be illustrated by examining a portion of the data from a peppermint oil analysis. Figure 2a shows the total ion chromatogram for a sample of a peppermint flavor extract. Some of the components are clearly unresolved with these chromatographic conditions. If we look at the mass spectrum of the component eluting at 15 minutes (Figure 2b), we can see a resolved doublet at m/z 139. The doublet is shown in an enlarged view of the region of the mass spectrum around m/z 139 (Figure 2c).

If we were to look at a single-mass chromatogram for mass 139 (+/- 0.5 amu), we would see only two clearly-defined chromatographic peaks (Figure 2d). But if we choose a more selective mass range to take advantage of the high resolution information, we can clearly identify several other components. In fact, three isobaric ions may be identified, two of which are associated with the sample, and one which is present in the background. The high-resolution single-ion reconstructions (Figure 3) clearly show which ion is associated with which eluting components.

The background ion is only identifiable from its suppression when components are eluting. This sort of phenomenon is known to occur in conventional GC/MS analyses; in our case the phenomenon may be associated with dynamic range considerations [17]. From analysis of this same mixture by MS/MS, [18], we have identified the $C_9H_{15}O^+$ peaks as fragments from menthol and menthone. The $C_{10}H_{19}^+$ peaks are assigned as fragments from menthyl acetate, menthyl valerate, and cadinene.

The use of the dual-cell FTMS to perform a typical target compound analysis may be illustrated by the identification of cocaine in a human urine extract. To determine the retention time of cocaine, a drug standard containing 10 ng of cocaine was injected using splitless injection onto a 20 meter BP-1 column with a 200 micron ID. The column was heated from 90 to 250 degrees at 10 degrees per minute. Under these conditions, the component eluting at 15.8 minutes (Figure 4a) was identified as cocaine. The mass spectrum corresponding to this peak (Figure 6a), shows the molecular ion at m/z 303, as well as characteristic fragment ions. The total ion chromatogram (Figure 4b) from the urine extract is quite complex and does not have a strong peak at 15.8 minutes. However, by plotting mass chromatograms (Figure 5) of the base peak in the mass spectrum of cocaine (m/z 82) for both the standard and the urine extract, one can see a large peak at an elution time of 15.8 minutes.

The mass spectrum corresponding to this peak (Figure 6b) matches well with the spectrum obtained from the drug standard, and also with the (Wiley NBS) library spectrum. Using the accurate mass measurements obtained on the FTMS data, comparisons can be made between the standard and the unknown. Based on twelve

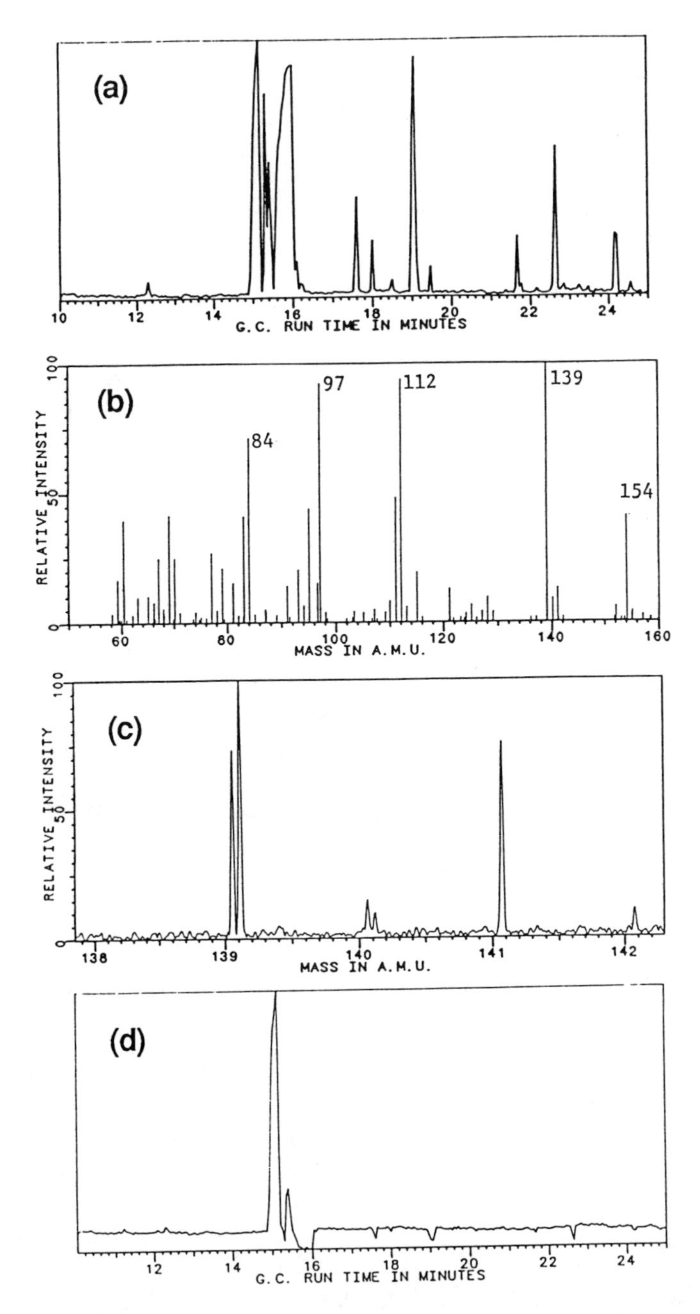


Figure 2. Peppermint oil analyzed by GC/FTMS: (a) total ion chromatogram; (b) mass spectrum of component eluting at 15 minutes; (c) enlargement of region around m/z 139 from (b); (d) mass chromatogram for m/z 139 +/- 0.5 amu.

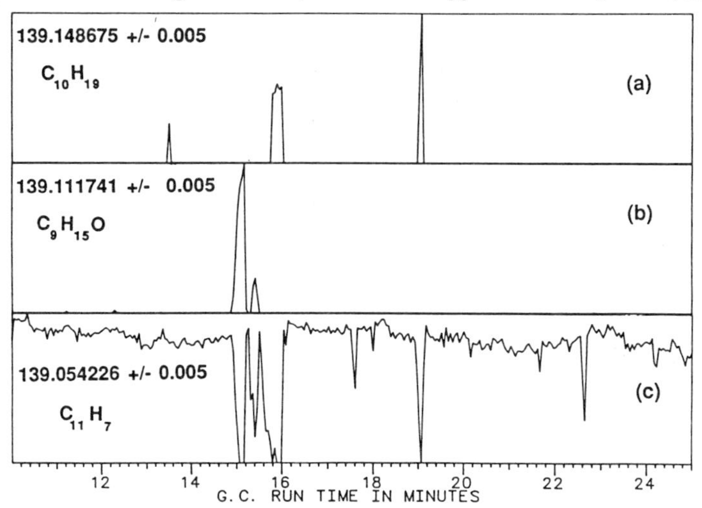


Figure 3. High resolution mass chromatograms for m/z 139: (a) mass chromatogram for $C_{10}H_{19}^+$; (b) mass chromatogram for $C_9H_{15}O^+$; (c) mass chromatogram for $C_{11}H_{17}^+$ (background peak).

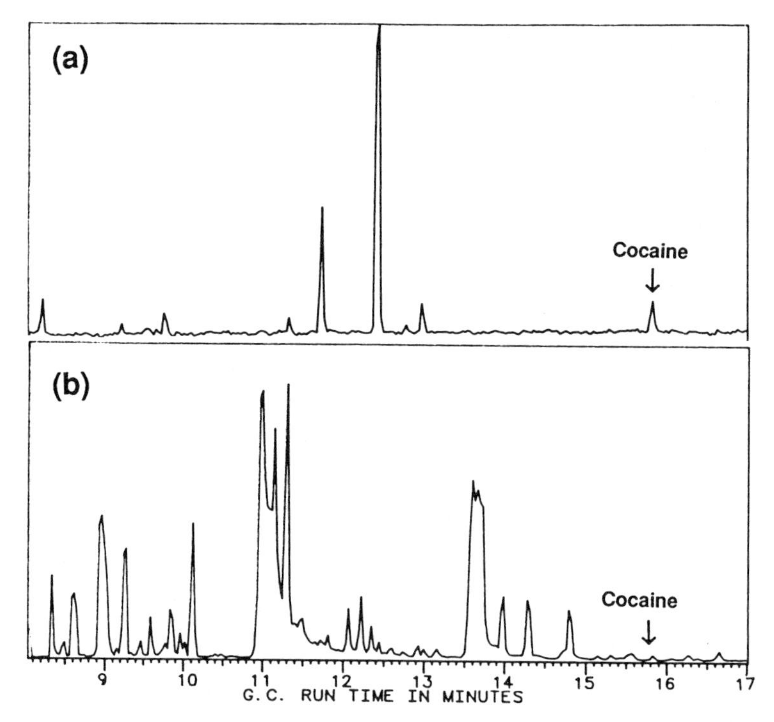


Figure 4. (a) chromatogram for drug standards including cocaine; (b) Chromatogram for urine extract.

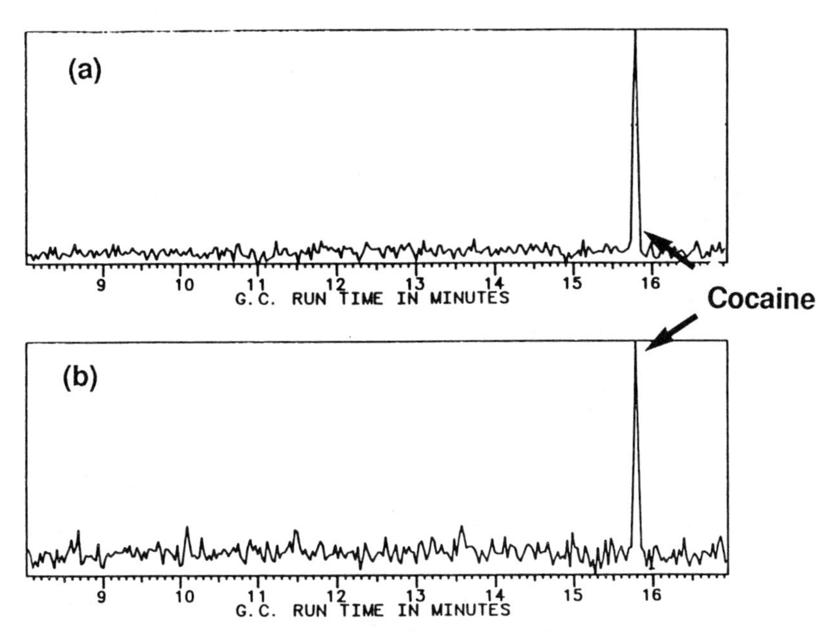


Figure 5. Mass chromatograms for m/z 82 in: (a) standard drug mixture; (b) cocaine.

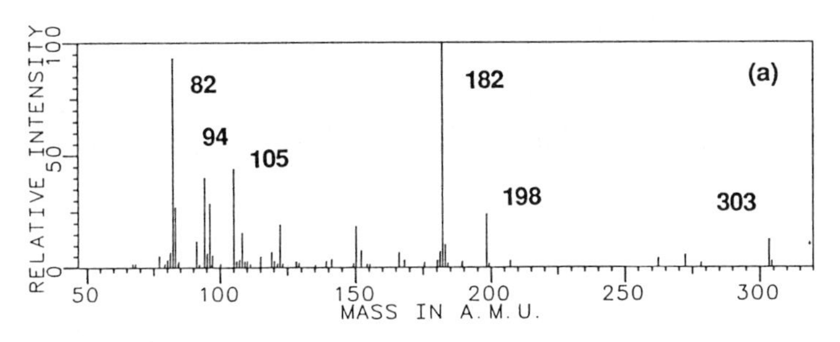


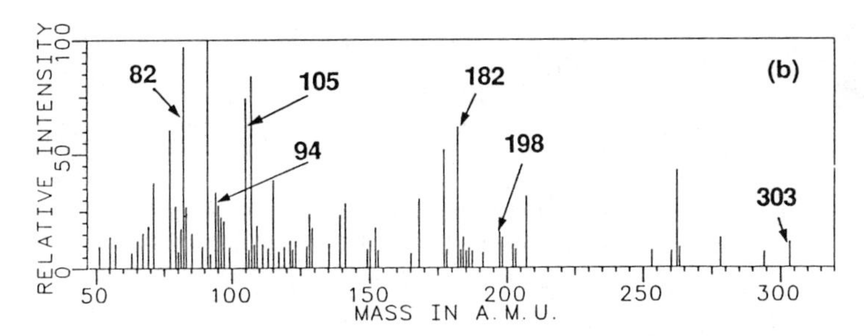


Figure 6. (a) Mass spectrum of cocaine standard; (b) Mass spectrum of component in urine extract eluting at 15 minutes (unsubtracted).

significant ions in the two spectra, mass measurement differences (based on 32K data points, and without internal calibrant) averaged only 8 ppm, adding an additional degree of confidence to the assignment of the unknown as cocaine.

Supercritical Fluid Chromatogaphy/FTMS. Supercritical fluid chromatography (SFC) has experienced a recent growth in popularity as an alternative approach to liquid chromatography, due to its increased resolution and faster analysis times. SFC is particularly attractive as a chromatographic method for coupling with FTMS, due to the reduced gas burden, in comparison with HPLC. We have performed preliminary experiments coupling SFC with the dual-cell Fourier transform mass spectrometer to demonstrate the capability of the dual-cell geometry to deal with the gas load imposed by SFC. These experiments indicate that both high mass spectrometric resolution and accurate mass measurements are possible with SFC/FTMS [19].

For this work, a 5 meter x 50 micron ID fused silica column, coated with a 0.25 micron polydimethylsiloxane film was introduced directly into the source chamber through the transfer line normally used for GC/FTMS. A restrictor was created at the end of the column by using a microflame to draw out the end of a 1 meter portion of deactivated but uncoated column to an inside diameter of approximately one micron. Details of the instrumentation used for SFC have been described elsewhere [19]. With the SFC interface in place, pressures in the source chamber were approximately 5 x $10^{-5}$ torr. Despite this high source cell pressure, we were able to obtain relatively high quality mass spectral data with analyzer side detection at 5 x $10^{-7}$ torr.

Figure 7a shows a total ion chromatogram of the separation of two model compounds, caffeine and methyl stearate, by SFC/FTMS. The isopentane CI spectra (Figures 7b and 7c) of each of these compounds shows only the $[M+H]^+$ peaks and the carbon 13 isotope peaks for each of the two compounds. Though crude, these results demonstrate the feasibility of coupling SFC with FTMS.

The high resolution obtainable for SFC/FTMS is shown in Figure 8, which shows the molecular ion region of caffeine, taken under self-CI conditions (no reagent gas). Separation of the molecular and pseudomolecular ions is shown at a resolution of 8,000. It must be emphasized that this resolution was obtained for a direct mode spectrum, over the mass range 50 amu up to the highest mass measured, using a data collect of 128K data points. This resolution is only data point limited, and does not reflect the maximum attainable at the working pressures of SFC/FTMS. We expect that resolutions in excess of 10,000 should be readily attainable [55]. Accurate mass measurements in the low ppm range for caffeine taken in the EI mode are shown in Table II.

Laser Desorption/FTMS. Laser desorption (LD)/FTMS [20] has been applied to a variety of chemical problems, ranging from fundamental studies of ion-molecule reactions [21,22], to the analysis of pharmaceuticals [4,23-29], to polymers [7,30-35], to organometallics [6-39], and to surface analysis [4,40]. The large number of papers being generated from the few laboratories working

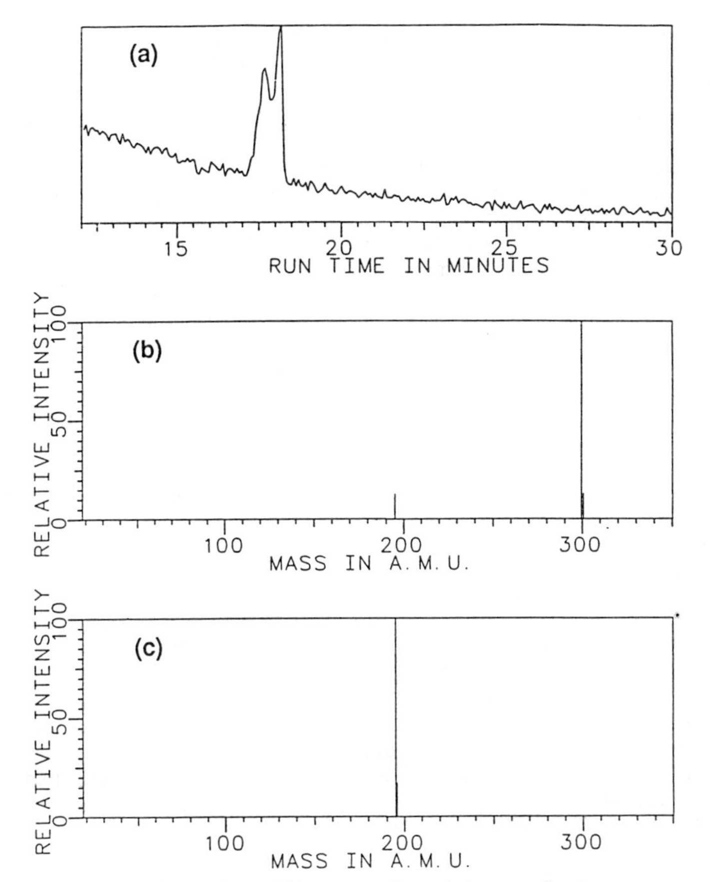


Figure 7. SFC/FTMS feasibility study: (a) total ion current for SFC/FTMS separation of methyl stearate; (b) isopentane CI mass spectrum of methyl stearate (with trace of caffeine); (c) isopentane CI mass spectrum of caffeine.

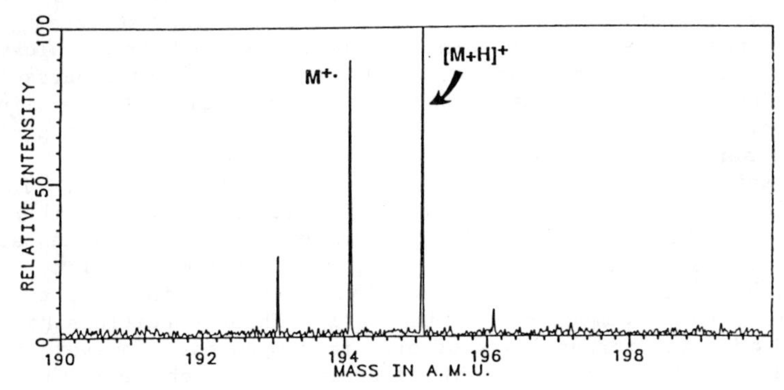


Figure 8. Molecular ion region of SFC/FTMS spectrum of caffeine under self-CI conditions, detected at a resolution of 8,300.

with LD/FTMS should serve to indicate the growing importance of the methods. In this article, we present two specific applications of LD/FTMS to surface and inorganic analysis.

A simple example of the use of LD/FTMS for surface analysis may be found in the analysis of blue stain on the surface of a copper part. In order to determine the nature of the stain, we obtained LD/FTMS spectra for samples of both stained and unstained copper. Both samples were mounted on the automatically rotated sample insertion probe on the FTMS-2000. The samples were rotated after each laser shot to expose a fresh surface at the laser focus. Approximately twenty-five laser shots were signal averaged for each spectrum, in order to increase signal-to-noise, and to provide a spectrum which would represent the averaged composition of the surface.

This approach was employed, since spectra of single spots on the surface might have local variations in composition which would hinder a comparison of the two samples. The spot size at the focal point was approximately 150 microns in diameter. The instrument was operated in dual-cell mode, with ions detected in the analyzer side immediately after the laser was fired, in order to minimize undesirable ion-molecule reactions which might occur at the higher pressures in the source cell. Both positive and negative ion spectra were collected for each sample.

Positive ion spectra were not particularly helpful, showing only $Cu^+$ and trace alkali metal ions ($Na^+$, $K^+$). However, the negative ion spectra were more complex, and a comparison between the spectrum of the stained copper (Figure 9a) with the unstained copper (Figure 9b) shows several additional peaks in the spectrum of the stained copper. Examination of these peaks revealed them to result from contamination of the surface by sulfur, with prominent peaks identified as:

$$S^-,\ HS^-,\ S_2^-,\ CuS_2^-,\ CuS_2H^-,\ Cu_2S_2^-,\ Cu_2S_2H^-,\ \text{and}\ Cu_3S_3^-$$

These ions were identified by mass measurement as well as their isotopic abundances. Resolution for these spectra was approximately 5,000 which was adequate to separate $CuS_2^-$ from traces of $I^-$ present on the surface as a contaminant in both samples. The analysis was confirmed by Auger spectroscopy, which clearly showed the presence of sulfur on the surface. The contamination was later found to have resulted from vulcanized rubber that had come in contact with the stained copper parts.

Another example is found in the analysis of the mineral zircon. We had previously published [4] a spectrum of a positive ion laser desorption spectrum of a sample of the mineral zircon (zirconium silicate) showing uranium as $^{238}U^+$, present in the sample at a level of approximately 15 parts-per-million [41]. The spectrum, which showed mixed zirconium oxides and hydroxides as the most intense peaks in the spectrum, was taken with a four second delay between the laser pulse and ion detection, in order to allow neutrals to be pumped out of the cell. These conditions had been found adequate for analysis of organic compounds. However, it was found that the reactivity of zirconium was such that the mixed oxides and hydroxides were produced as ion-molecule reaction products during the long trap period.

Table II. SFC-FTMS, Electron Impact Measured Masses for Caffeine

| COMPOSITION | CALCULATED | MEASURED | DIFFERENCE (ppm) |
|---|---|---|---|
| $C_8H_{10}N_4O_2^+$ | 194.080376 | 194.080307 | -0.36 |
| $C_8H_9N_4O_2^+$ | 193.072551 | 193.072597 | -0.23 |
| $C_4H_6N_2^+$ | 82.053098 | 82.052900 | -2.4 |

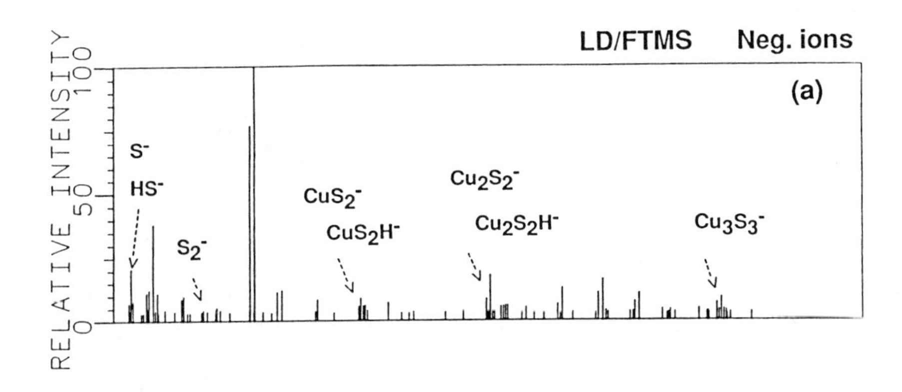


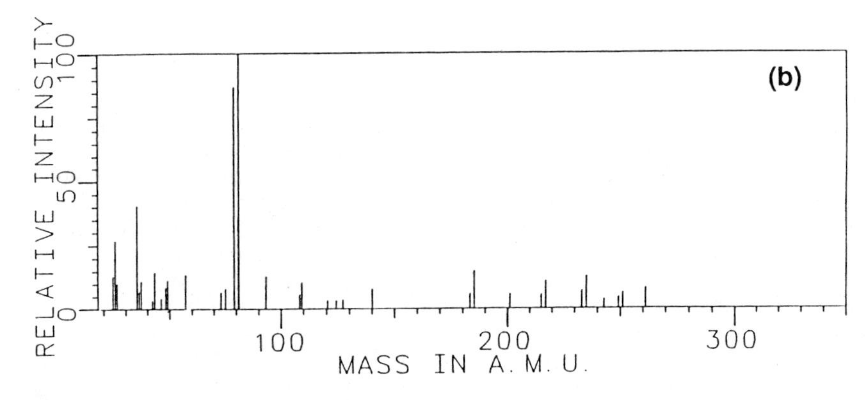


Figure 9. Analysis of a surface stain on copper showing copper sulfide contamination: (a) stained copper; (b) "clean" copper standard.

By reducing the delay time from four seconds to 100 milliseconds, we find that the ion-molecule reaction products are significantly reduced, and that the base peak is now due to $Zr^+$. A laser desorption spectrum of a sample of zircon containing approximately 100 times more uranium than the previous sample is shown in Figure 10. This spectrum shows the presence of lead resulting from the radioactive decay of uranium, as well as uranium and uranium oxide peaks. Trace rare earths are also evident, giving peaks due to $Er^+$, $Yb^+$, $Y^+$, and $Hf^+$. Unlike the previously published example, no ejection was required in order to see the uranium. The ratio of $^{238}U$ to its radioactive decay product, $^{206}Pb$, may be used to date the mineral. The ratios for this sample gave a calculated age of 1.5 million years, which agreed with the age determined by standard geological methods.

MS/MS. The capability of trapping ions for long periods of time is one of the most interesting features of FTMS, and it is this capability that has made FTMS (and its precursor, ion cyclotron resonance) the method of choice for ion-molecule reaction studies. It is this capability that has also lead to the development of MS/MS techniques for FTMS [11]. FTMS has demonstrated capabilities for high resolution daughter ion detection [42-44], and consecutive MS/MS reactions [45], that have shown it to be an intriguing alternative to the use of the instruments with multiple analysis stages. Initial concerns about limited resolution for parent ion selection have been allayed by the development of a stored waveform, inverse Fourier transform method of excitation by Marshall and coworkers [9,10] which allows the operator to tailor the excitation waveform to the desired experiment.

In addition, several approaches to ion activation that are not well-suited to conventional types of mass spectrometers, are very well-suited for use with FTMS. These include photodissociation and ion-electron collisions [46-49]. In this paper, we would like to present some applications of ion-electron collisions to MS/MS, and show that the method may be a suitable alternative to collisional activation for some applications. We will also discuss some of the desirable characteristics of the dual cell geometry, as applied to MS/MS experiments.

Positive ions may be trapped by electrostatic attraction in the ionizing electron beam, if the electron beam current is sufficiently high [50]. Ion-electron collisions may cause enough energy to be deposited in the trapped ions to result in further ionization [51], or activation resulting in bond breaking [52]. Activation of ions by ion-electron collisions results in fragmentation which may be compared to collisional activation, and which may be used to obtain breakdown curves [53]. The ion-electron collisional activation approach has been given the tongue-in-cheek name, "Electron Impact Excitation of Ions from Organics", or, "EIEIO".

We have recently shown that it is possible to perform MS/MS experiments by making use of the ion-electron collisions in a Fourier transform mass spectrometer [54]. Parent ion selection, which could not easily be achieved in the previous experiments, is accomplished by gating off the electron beam for a short period of time, in order to permit ion ejection. The high field of the superconducting magnet helps keep the ions from drifting out of the electron beam, allowing them to undergo collisions with electrons when the beam is gated on again.

An example is shown in Figure 11, which shows the method applied to a synthetic mixture, containing acetyl acetone, N-methyl aniline, isophorone, and acetophenone. After ions have been formed by electron impact, they are confined in the source cell and allowed to undergo ion-molecule reactions ("self-CI") for a period of 100 milliseconds (Figure 11a). The electron beam is left on during this reaction period at a low electron energy (3 eV) in order to help confine ions in the electron beam. Following the reaction period, the electron beam is gated off for approximately 3-5 milliseconds, during which two swept-frequency ion ejection events select out the $[M+H]^+$ ion (and some molecular ion) for acetyl acetone, which has a relative abundance of only 1-2 percent due to the high proton affinity of the N-methyl aniline. The electron beam is then gated back on, and the conductance limit is grounded for a period corresponding to one trapping oscillation for the ions. This has the effect of transferring virtually all of the $[M+H]^+$ ions into the analyzer cell (Figure 11b). The ions remain trapped in the electron beam for a period of 100 milliseconds, at a low electron energy, (4 eV) and a high electron current (25 microamperers). During this period, the ions undergo activation and fragmentation, resulting on the formation of daughter ions at m/z 42 and m/z 85 (Figure 11c). Selection of the $M^+$ and $[M+H]^+$ ions for N-methyl aniline and the EIEIO spectrum for this component is shown in Figures 12a and 12b.

In addition to the EIEIO experiments, the dual cell has also been employed for collisional activation (CA) experiments. Selected parent ions may be transferred to the analyzer cell for CA. With pulsed valve introduction of the collision gas (argon) into the analyzer cell, it is possible to obtain very high daughter ion resolution. Figure 13 shows a $C_6H_5CO^+/C_8H_9{}^+$ doublet from CA of acetophenone and mesitylene, detected at a resolution of 500,000.

A major difficulty in using FTMS for complex mixture analysis by MS/MS techniques has been the tendency of daughter ions to react with neutral species from the sample. These unwanted ion-molecule reactions can complicate the interpretation of MS/MS spectra, or even prevent the observation of daughter ions altogether. However, we may use the dual-cell geometry to transfer selected parent ions into the analyzer cell, where they are isolated from reactive neutral species. If this is done, then the use of FTMS with collisional activated dissociation (CAD) for complex mixture analysis is greatly facilitated. As an example, we consider the determination of which of three possible isomers is present in an epoxy resin extract. For one component of the extract, diaminodiphenylsulfone (DADPS), it has been shown using a hybrid instrument of BEQQ geometry [55] that CAD spectra are substantially different for different isomers.

We have repeated this work to demonstrate the use of the dual-cell FTMS to distinguish isomers in a mixture. Figure 14a shows an electron impact spectrum of the epoxy resin extract introduced via the direct insertion probe. Figure 14b shows the 50 eV CAD

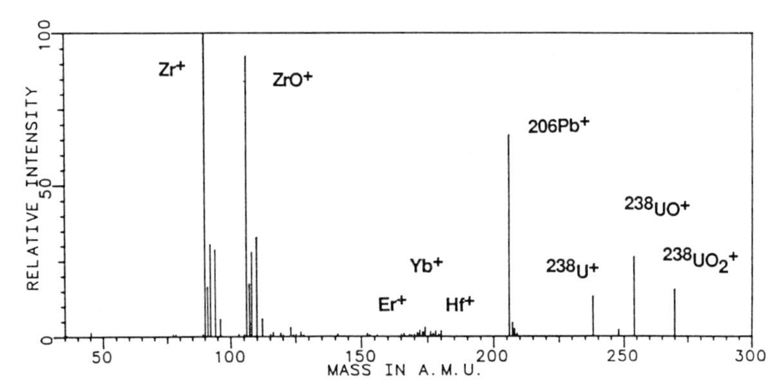


Figure 10. Mass Spectrum of zircon mineral (sample courtesy of M. Harrison and S. deLong, SUNY, Albany).

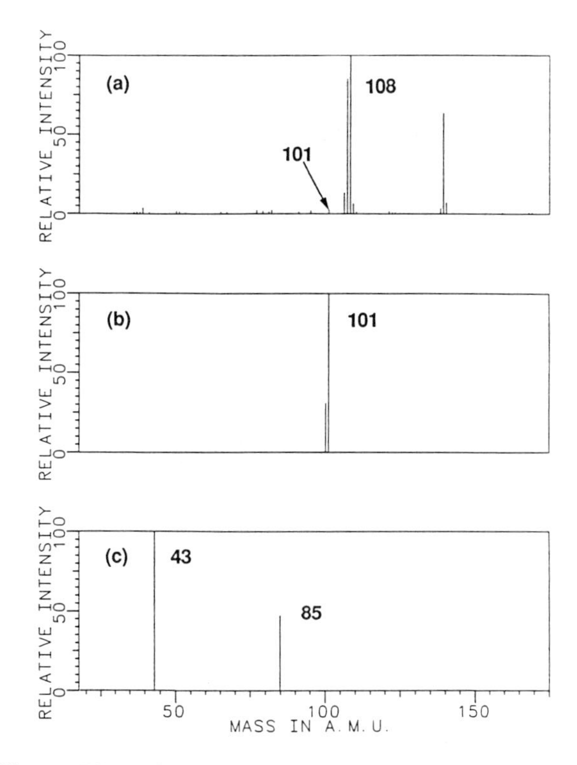


Figure 11. Identification of acetyl acetone in a synthetic mixture by EIEIO: (a) Self CI mass spectrum showing acetyl acetone at .2%; (b) Selection of acetyl acetone ($M^+$ and $M+1^+$); (c) EIEIO fragments from acetyl acetone in mixture.

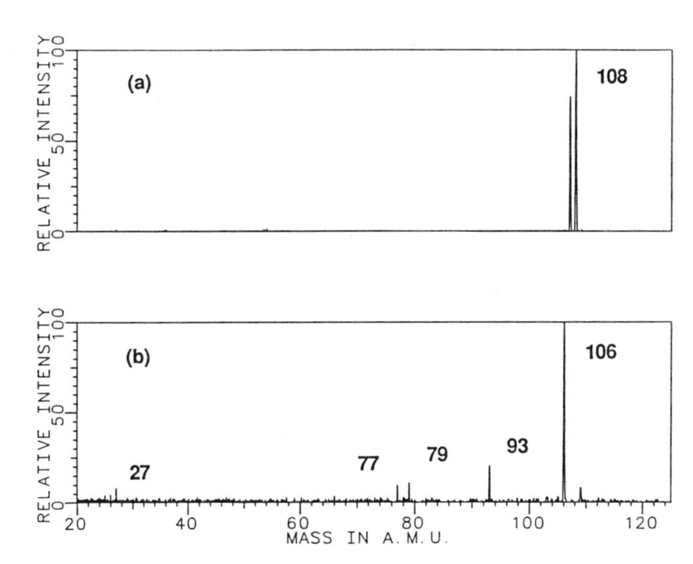


Figure 12. Identification of N-methyl aniline in the synthetic mixture by EIEIO: (a) Selection of N-methyl aniline; (b) EIEIO fragments of N-methyl aniline in the mixture.

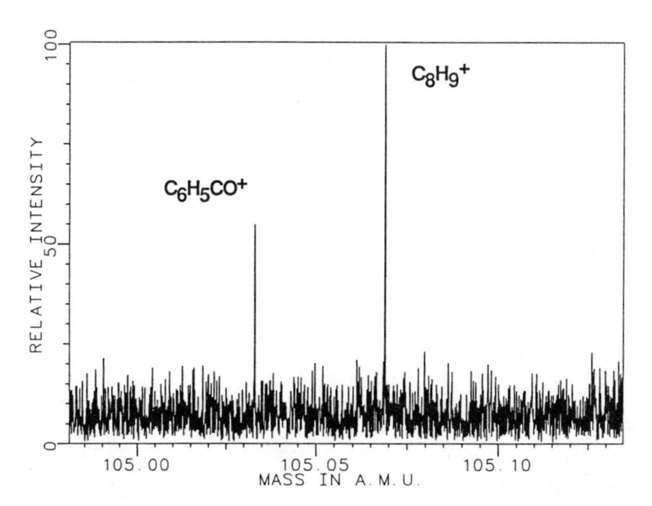


Figure 13. Isobaric daughter ions from CAD of acetophenone and mesitylene, detected at a resolution of 500,000.

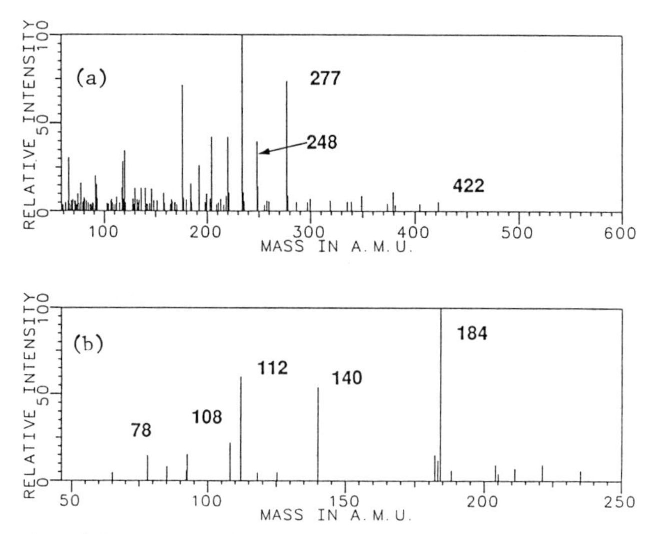


Figure 14. (a) EI spectrum of epoxy resin extract: (b) 50 eV CAD spectrum of m/z 248 component in epoxy resin extract.

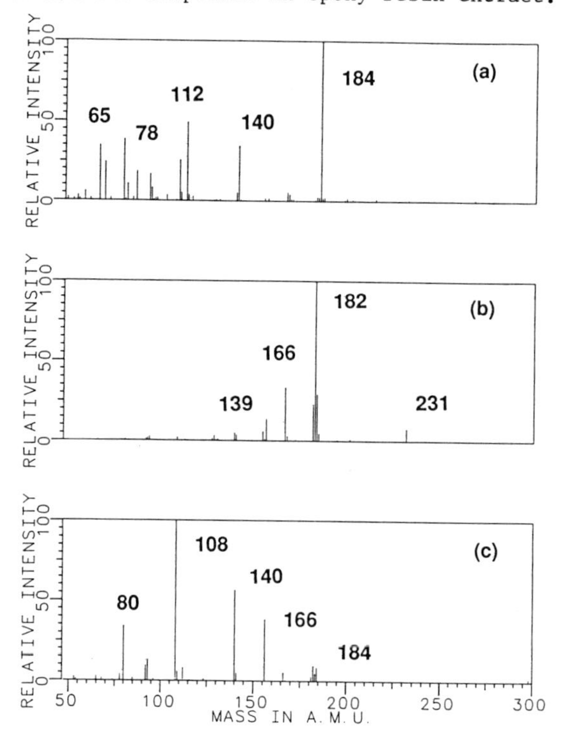


Figure 15. 50 eV CAD spectra of pure diaminodiphenylsulfone isomers: (a) 3,3'-diaminodiphenylsulfone; (b) 2,4'-diaminodiphenylsulfone; (c) 4,4'-diaminodiphenylsulfone.

spectrum of the m/z 248 ion, using argon as the collision gas. A comparison of this spectrum with the 50 eV CAD spectra obtained for three different isomers of DADPS (Figures 15a, 15b, and 15c) clearly shows that the component in the epoxy resin extract is the 3,3′ isomer. While there are some differences observed in comparing the FTMS-CAD spectra to those reported for the hybrid instrument, this work verifies that FTMS-CAD provides a viable approach to this type of analysis.

## Conclusion

The dual cell geometry has been employed to overcome difficulties resulting from high gas loads associated with coupling both gas chromatography and supercritical fluid chromatography with FTMS. For laser desorption and MS/MS experiments, the dual cell geometry has allowed us to separate ions from reactive neutrals, thus eliminating unwanted ion-molecule reactions. These developments have enabled us to apply FTMS to a variety of analytical problems, as evident from the examples in this report.

## Acknowledgments

We would like to thank Jim Scrivens of ICI, Ltd. for the epoxy resin extract and guidance on the use of CAD for isomer differentiation in the mixture.

We would also like to extend our appreciation to Jack Henion and Ed Lee of Cornell for their collaboration on the SFC work. The authors are grateful to Gail Harris for her assistance with the manuscript.

## Literature Cited

1. M. B. Comisarow and A. G. Marshall, "Frequency Sweep Fourier Transform Ion Cyclotron Resonance Spectroscopy," Chem. Phys. Lett., 26, 489-490 (1974).
2. D. F. Hunt, J. Shabanowitz, R. T. McIver, Jr., R. L. Hunter, and J. E. P. Syka, "Ionization and Mass Analysis of Nonvolatile Compounds by Particle Bombardment Tandem-Quadrupole Fourier Transform Mass Spectrometry," Anal. Chem., 57, 765-768 (1985).
3. U.S. Patent Application, Serial No. 610502.
4. R. B. Cody, J. A. Kinsinger, Sahba Ghaderi, I. J. Amster, F. W. McLafferty, and C. E. Brown, "Developments in Analytical Fourier Transform Mass Spectrometry," Analytica Chimica Acta, 178, 43-66 (1985).
5. P. Kofel, M. Allemann, H. P. Kellerhals, and K. P. Wanczek, "External Generation of Ions in ICR Spectrometry, " Int. J. Mass Spec. Ion Proc., 65, 97-103 (1985).
6. I. J. Amster, F. W. McLafferty, M. E. Castro, D. H. Russell, R. B. Cody, and S. Ghaderi, "Detection of Mass 16,241 Ions by Fourier-Transform Mass Spectrometry," Anal. Chem., 58, 483-485 (1986).

7. C. L. Wilkins, D. A. Weil, C. L. C. Yang, and C. F. Ijames, "High Mass Analysis by Laser Desorption Fourier Transform Mass Spectrometry," Anal. Chem., 57, 520-524 (1980).
8. J. Shabanowitz, D. F. Hunt, M. Castro, and D. H. Russell, presented at the 34th Annual Conference on Mass Spectrometry and Allied Topics, Cincinnati, OH, June 8-13, 1986.
9. A. G. Marshall, T.-C. L. Wang, and T. L. Ricca, "Tailored Excitation for Fourier Transform Ion Cyclotron Resonance Spectrometry," J. Am. Chem. Soc., 107, 7893-7897 (1985).
10. T.-C. L. Wang, T. L. Ricca, and A. G. Marshall, "Extension of Dynamic Range in Fourier Transform Ion Cyclotron Resonance Mass Spectrometry via Stored Waveform Inverse Fourier Transform Excitation," Anal. Chem., 58, 2935-2938.
11. R. B. Cody, R. C. Burnier, and B. S. Freiser, "Collision-Induced Dissociation with Fourier Transform Mass Spectrometry," Anal. Chem., 54, 96-101 (1982).
12. R. L. Settine, J. A. Kinsinger, and S. Ghaderi, "Fourier Transform Mass Spectrometry and Its Combination with High Resolution Gas Chromatography," European-Spectroscopy News, 58, 16-18 (1985).
13. M. Comisarow and A. G. Marshall, "Fourier Transform Cyclotron Resonance Spectroscopy," Chem. Phys. Lett, 25, 28-283 (1974).
14. E. B. Ledford, Jr., D. L. Rempel, and M. L. Gross, "Space Charge Effects in Fourier Transform Mass Spectrometry. Mass Calibration," Anal. Chem., 56, 2744-2748 (1984).
15. C. L. Johlman, D. A. Laude, Jr., and C. L. Wilkins, "Accurate Mass Measurement in the Absence of Calibrant for Capillary Column Gas Chromatography/Fourier Transform Mass Spectrometry," Anal. Chem., 57, 1040-1044 (1985).
16. R. L. Settine and R. B. Cody, unpublished results.
17. In other words, since the dynamic range of the system is about $10^3$ to $10^4$, the less abundant background ions may not be detected in the presence of an eluting component.
18. R. B. Cody, Anal. Chem., submitted for publication.
19. E. D. Lee, J. D. Henion, R. B. Cody, and J. A. Kinsinger, "Supercritical Fluid Chromatography/Fourier Transform Mass Spectrometry," Anal. Chem., 59, 1309-1312 (1987).
20. D. A. McCrery, E. B. Ledford, Jr., and M. L. Gross, "Laser Desorption Fourier Transform Mass Spectrometry," Anal. Chem., 54, 1435-1437 (1982).
21. R. C. Burnier, G. D. Byrd, T. J. Carlin, M. B. Wise, R. B. Cody, and B. S. Freiser, "Study of Atomic Metal Ions Generated by Laser Ionization," pp. 98-118 in Ion Cyclotron Resonance Spectrometry 2, ed. K.-P. Wanczek, Springer-Verlag, West Germany (1982).
22. B. S. Freiser, "Applications of Laser Ionization/Fourier-Transform Mass Spectrometry to the Study of Metal Ions and Their Clusters in the Gas Phase," Anal. Chim. Acta., 178 (1), 137-158 (1985).
23. R. E. Hein and R. B. Cody, "Laser Desorption Fourier Transform Mass Spectrometry of Organic Compounds," Presented at the 31st Annual Conference on Mass Spectrometry and Allied Topics, Boston, MA (May 8-13, 1983).

24. R. E. Shomo, II, A. G. Marshall, and C. R. Weisenberger, "Laser Desorption Fourier Transform Ion Cyclotron Resonance Mass Spectrometry vs. Fast Atom Bombardment Magnetic Sector Mass Spectrometry for Drug Analysis," Anal. Chem., 57, 2940-2944 (1985).
25. C. E. Brown, S. C. Roerig, V. T. Berger, R. B. Cody, and J. M. Fujimoto, "Analgesic Potencies of Morphine 3- and 6-Sulfates After Intracerebroventricular Adminstration in Mice: Relationship to Structural Characteristics Defined by Mass Spectrometry and Nuclear Magnetic Resonance," J. Pharm. Sci., 74, 821 (1985).
26. M. L. Coates and C. L. Wilkins, "Laser Desorption Fourier Transform Mass Spectra of Malto-Oligosaccarides," Biomed. Mass Spec., 12, 424-428 (1985).
27. M. L. Coates and C. L. Wilkins, "Laser Desorption/Fourier-Transform Mass Spectrometry Mass Spectra of Glycoalkaloids and Steroid Glycosides," Biomed. and Env. Mass Spectrom., 13, 19-204 (1986).
28. D. A. McCrery and M. L. Gross, "Laser Desorption/Fourier-Transform Mass Spectrometry for the Study of Nucleosides, Oligosaccharides, and Glycosides," Anal. Chim. Acta., 178 (1), 91-103 (1985).
29. K. Faull and R. B. Cody, unpublished results.
30. C. E. Brown, P. Kovacic, C. A. Wilkie, R. B. Cody, and J. A. Kinsinger, J. Polymer Sci.: Polymer Lett. Ed., 23, 453 (1985).
31. C. E. Brown, P. Kovacic, C. A. Wilkie, J. A. Kinsinger, R. E. Hein, S. I. Yaniger, and R. B. Cody, J. Polymer Sci.: Polymer Chem. Ed., 24, 255-267 (1986).
32. C. A. Wilkie, J. Smukalla, R. B. Cody, and J. A. Kinsinger, J. Polymer Sci.: Part A: Polymer Chemistry, 24, 1297-1311 (1986).
33. C. E. Brown, P. Kovacic, C. A. Wilkie, R. B. Cody, R. E. Hein, and J. A. Kinsinger, "Laser Desorption/Fourier Transform Mass Spectral Analysis of Various Conducting Polymers," Synthetic Metals, 15, 265-279 (1986).
34. C. E. Brown, P. Kovacic, R. B. Cody, Jr., R. E. Hein, and J. A. Kinsinger, J. Polymer Sci.: Part C: Polymer Letters, 24, 519-528 (1986).
35. R. S. Brown, D. A. Weil, and C. L. Wilkins, "Laser Desorption-Fourier Transform Mass Spectrometry for the Characterization of Polymers," Macromolecules, 19, 1255-1260 (1986).
36. R. R. Weller, J. R. Eyler, and C. M. Riley, "Fourier Transform Mass Spectrometry of Cisplatin," J. Pharm. & Biomed. Anal., 3, 87-94 (1985).
37. M. B. Comisarow, D. P. Fryzuk, and R. B. Cody, unpublished results.
38. S. G. Shore, D.-Y. Jan, W.-L. Hsu, S. Kennedy, J. C. Huffman, T.-C. L. Wang, and A. G. Marshall, J. Chem. Soc. Chem. Comm., 392-394 (1984).
39. L.-Y. Hsu, W.-L. Hsu, and D.-Y. Jan, "S. G. Shore," Organometallics, 3, 591-595 (1984).

40. M. G. Sherman, J. R. Kingsley, J. C. Hemminger, and R. T. McIver, Jr., "Surface Analysis by Laser Desorption of Neutral Molecules with Detection by Fourier-Transform Mass Spectrometry," Anal. Chim. Acta., 178 (1), 79-89 (1985).
41. The concentration reported in the previous publication was in error.
42. R. B. Cody and B. S. Freiser, "High Resolution Detection of Collision-Induced Dissociation Fragments by Fourier Transform Mass Spectrometry," Anal. Chem., 54, 1431-1433 (1982).
43. R. L. White and C. L. Wilkins, "Low-Pressure Collision-Induced Dissociation Analysis of Complex Mixtures by Fourier Transform Mass Spectrometry," Anal. Chem., 54, 2211-2215 (1982).
44. T. J. Carlin and B. S. Freiser, "Pulsed Valve Addition of Collision and Reagent Gases in Fourier Transform Mass Spectrometry," Anal. Chem., 55, 571-574 (1983).
45. R. B. Cody, R. C. Burnier, D. J. Cassady, and B. S. Freiser, "Consecutive Collision-Induced Dissociation in Fourier Transform Mass Spectrometry," Anal. Chem., 54, 2225-2228 (1982).
46. S. A. McLuckey, L. Sallans, R. B. Cody, R. C. Burnier, S. Verma, B. S. Freiser, and R. G. Cooks, Int. J. Mass Spectrom. Ion Proc., 44, 215-219 (1982).
47. R. B. Cody and B. S. Freiser, Anal. Chem., 51, 547-551 (1979).
48. D. M. Fedor, R. B. Cody, D. J. Burinsky, B. S. Freiser, and R. G. Cooks, Int. J. Mass Spectrom. Ion Proc., 39, 55-64 (1981).
49. B. S. Freiser, Anal. Chem., 59, 1054-1056 (1987).
50. B. S. Freiser, "Trapping of Positive Ions in the Electron Beam of an Ion Cyclotron Resonance Spectrometer," Int. J. Mass Spectrom. Ion Proc., 26, 39-47 (1978).
51. B. S. Freiser, "Electron Impact Ionization of Argon Ions by Trapped Ion Cyclotron Resonance Spectrometry," Int. J. Mass Spectrom. Ion Proc., 33, 263-267 (1980).
52. R. B. Cody and B. S. Freiser, "Electron Impact Excitation of Ions from Organics: An Alternative to Collision Induced Dissociation," Anal. Chem., 51, 547-551 (1979).
53. D. M. Fedor, R. B. Cody, D. J. Burinsky, B. S. Freiser, and R. G. Cooks, "Dissociative Excitation of Gas-Phase Ions. A. Comparison of Techniques Utilizing Ion Cyclotron Resonance Spectroscopy and Angle-Resolved Mass Spectrometry," Int. J. Mass Spectrom. Ion Proc., 39, 55-64 (1981).
54. R. B. Cody and B. S. Freiser, "Electron Impact Excitation of Ions in Fourier Transform Mass Spectrometry," Anal. Chem., 59, 1054-1056 (1987).
55. T. G. Blease, J. H. Scrivens, D. A. Catlow, E. Clayton, and J. J. Monaghan, Int. J. Mass Spec. Ion Proc., submitted.

RECEIVED September 9, 1987

Chapter 5

# Instrumentation and Application Examples in Analytical Fourier Transform Mass Spectrometry

**Frank H. Laukien [1], M. Allemann [2], P. Bischofberger [2], P. Grossmann [2], Hp. Kellerhals [2], and P. Kofel [2]**

**[1]Department of Chemical Physics, Harvard University, Cambridge, MA 02138**
**[2]Spectrospin AG, Industriestrasse 26, CH-8117 Fallanden, Switzerland**

Selected topics in Fourier-Transform Ion Cyclotron Resonance Mass Spectrometry instrumentation are discussed in depth, and numerous analytical application examples are given. In particular, optimization ofthe single-cell FTMS design and some of its analytical applications, like pulsed-valve CI and CID, static SIMS, and ion clustering reactions are described. Magnet requirements and the software used in advanced FTICR mass spectrometers are considered. Implementation and advantages of an external differentially-pumped ion source for LD, GC/MS, liquid SIMS, FAB and LC/MS are discussed in detail, and an attempt is made to anticipate future developments in FTMS instrumentation.

It is not the intention of this chapter to provide a comprehensive treatment of FTMS technology and instrumentation, since excellent reviews exist (1). Instead, the focus will be on selected instrumental requirements which are crucial for the performance of a modern FTICR mass spectrometer, and which are not usually discussed elsewhere. Rather than striving for comprehensiveness, this paper is written with the intention of discussing specific design aspects in depth; the emphasis is on the unique features of our mass spectrometer, since other instrumental aspects are elaborated on elsewhere in this book.

Specifically, four instrumental topics are covered: First, our design criteria for an optimized single cell are explained and illustrated by several analytical application examples, including pulsed-valve CI and CID, static SIMS, and gas-phase polymerization reactions. Second, we explore the magnet requirements of FTMS as far as field strength, homogeneity and stability are concerned. Third, we offer our design philosophy for FTMS applications software, and the rationale for some unique features like phasing, the provisions for rapid scan/correlation ICR, timesharing, and versatile experimental event sequencing. Fourth, we discuss and illustrate

[1]Correspondence should be addressed to this author.

0097-6156/87/0359-0081$06.00/0

with analytical examples the desirability of a differentially-pumped ion source in general, and our implementation of external ionization in particular. Finally, an outlook on the anticipated future development of FTMS instrumentation and analytical techniques is provided.

## Single-Cell FTMS: Design and Applications

**ICR Cell Design.** The exact geometry of the ICR cell is experimentally not critical, except possibly in ultra-high resolution measurements. The experimental fact that in terms of both resolution and mass accuracy, elongated rectangular, cylindrical, and Penning traps are more or less equivalent, has been corroborated theoretically (2). The choice of geometry is therefore determined by the desirability of maximizing the cell diameter and volume for a given magnet and vacuum system, while retaining good resolution and undistorted line-shapes. For the cylindrically symmetric environment near the magnetic center of a supercon magnet, a cylindrical cell, as depicted in Figure 1, maximizes the cell diameter $d_x$ perpendicular to B. The volume of the cell employed in the CMS 47 is 170 ccm, and $d_x = d_z = 60$ mm.

If elongated cells ($d_z > d_x$) are utilized, we experimentally observe a degradation of both resolution and line-shape, presumably because the length of the homogeneous magnetic field region is approximately 60 mm.

The use of large cells in FTMS is advantageous, firstly because the volume of the quadrupolar electric field region is maximized and line-broadenings are therefore reduced, and secondly because the maximum trapping time increases with $d_{x2}$. This last point is particularly important if one is interested in large ion clusters which build up via gas-phase ion-neutral reactions. The average number of ion-neutral collisions C (and hence the maximum cluster size) which can occur before the ion clusters are lost at the side walls is proportional to the square of the product of magnetic field strength B and cell diameter $d_x$ according to (3)

$$C \propto (d_x B)^2 / V \qquad (1).$$

Here V is the trapping voltage; note that C is pressure-independent.

An example of large negative ion clusters building up from smaller parent ions in the ICR cell is given in Figure 2a, where the $Re_{17}$- and $Re_{18}$-groups are selected from the negative ion-cluster mass spectrum of decacarbonyldirhenium. After electron capture and loss of one or several carbonyl groups, negative rheniumcarbonyl clusters polymerize up to $Re^1{}_8(CO)_{31}^-$ at mass M = 4,216 amu; they can still be observed with unit resolution. Positive rheniumcarbonyl clusters up to M = 8,828 amu are plotted in Figure 2b. The average number of collisions experienced by the parent ion of mass 652 amu is approximately C = 15,000. The average collision number for smaller ions, such as water, can be as high as 400,000.

All broadband spectra were amplified by a low-noise preamplifier with a dynamic range > 1,000, and a broadband main amplifier. The amplified time-domain signal is digitized by a 20 MHz, 9 bit Bruker ADC with 128 K words of buffer memory, and Fourier transformed by a Bruker array processor (128 K word FFT in 8 sec).

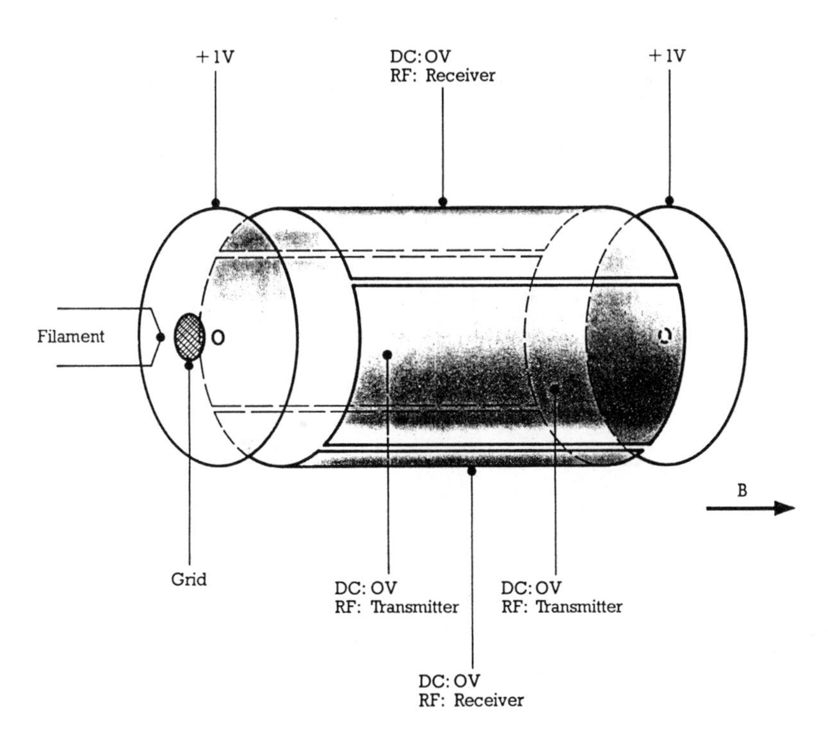


Figure 1. Cylindrical ICR cell with $d_x = d_z = 60mm$. The voltages indicated are suitable for positive ions and can simply be reversed for the study of negative ions.

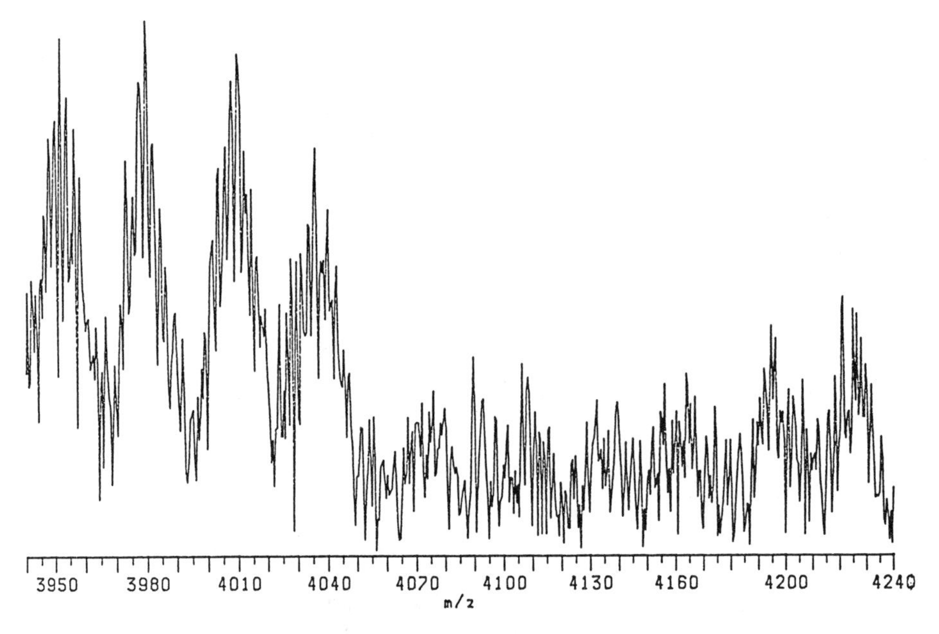


Figure 2a. Negative rheniumcarbonyl clusters with unit resolution above 4,200 amu ($2x10^{-8}$ mbar, 65 eV EI, CID activation, 1,000 scans).

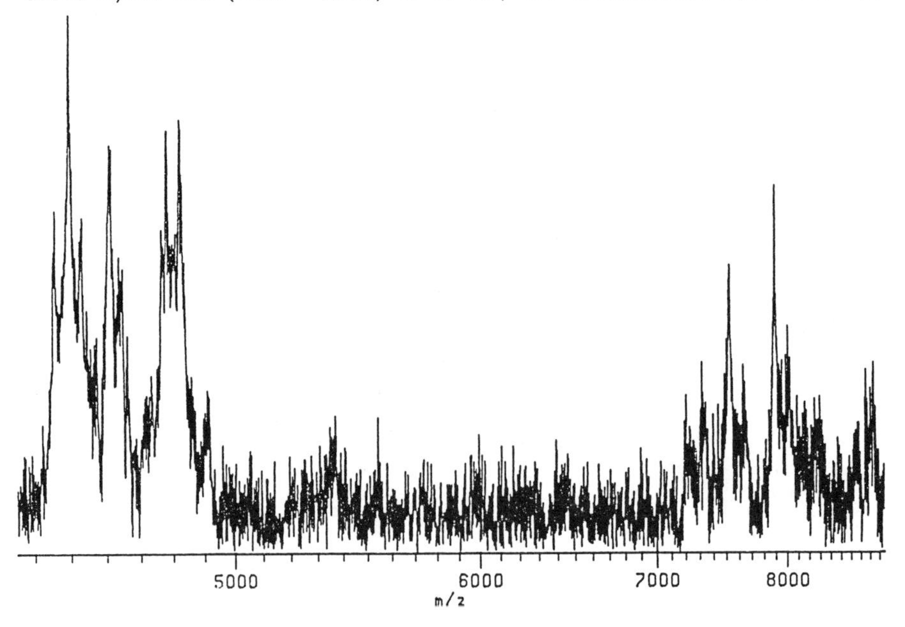


Figure 2b. Positive rheniumcarbonyl clusters with highest mass peaks > 8,800 amu ($2x10^{-8}$ mbar, 5000 scans).

The advantages of differential ion acceleration and detection in terms of rf power and signal homogeneity have been elucidated by Dunbar (4). In our cell design, rf voltages of up to 400 Volt (peak-to-peak) can be applied differentially, which allows for rapid acceleration and ejection.

Pulsed-Valve CI and CID-Experiments. Chemical Ionization (CI), self-CI (SCI), and direct or desorption CI (DCI) experiments in FTMS can be done equally well with the differentially-pumped external ion source described below, or with a pulsed-valve single cell arrangement (5,6). In our experiments, we admit a pulse of reagent gas via a piezoelectric pulsed valve with a minimum opening time of about 2.5 ms (7). Unlike solenoid pulsed valves, the performance of piezoelectric pulsed valves is not disturbed by the strong magnetic field of 4.7 Tesla.

Immediately following this pulse of reagent gas, the pressure in the ICR cell temporarily climbs to between $10^{-5}$ to $10^{-6}$ mbar. During a time delay of 1 - 2 seconds before acceleration and detection, the ICR cell is pumped down to $2x10^{-8}$ mbar by a 330 l/s turbomolecular pump, permitting enhanced resolution, and accurate mass determinations without internal calibrants.

Examples of both, pulsed-valve positive ion CI, using methane and ammonia as reagent ions, and pulsed-valve negative ion CI, using a $N_2O/CH_4$ mixture, are shown in Figure 3 a,b. In both examples the molecular ion was not stable with respect to electron impact at 70 eV, but the CI spectra clearly show abundant quasi-molecular-ion peaks.

Piezoelectric pulsed-valve inlet systems are equally useful in collision-induced dissociation (CID) experiments (8) where the CID target gas (usually Argon) is pulsed, and subsequently pumped away to permit high-resolution, high-accuracy acquisition of FTMS spectra.

Recently, low pressure CI and SCI experiments, which take advantage of the long reaction times (typically 10 to 60 seconds) possible in an FTMS, have been demonstrated (9). Figure 4 exhibits low-pressure EI and DCI spectra of Riboflavin (Vitamin B2), taken without a pulsed valve. Only the DCI spectrum taken at $2x10^{-8}$ mbar contains the quasi-molecular ion at M = 377.

SIMS. Secondary Ion Mass Spectrometry is particularly suited for ionization of nonvolatile, polar, and thermally labile molecules. Liquid SIMS, using liquid glycerol matrices, is best done in the differentially-pumped external ion source, because matrix effects and the high vapor pressure of glycerol make liquid SIMS unsuitable for single cell low-pressure FTMS.

Static SIMS, however, which does not require any matrix, and demands only low primary beam intensities, does not deteriorate the ultra-high vacuum in the FTMS single cell. An example of a static SIMS spectrum of Valinomycin using primary 2-4 kV $Cs^+$-ions is shown in Figure 5, including a high-resolution spectrum of the quasi-molecular (M+Na)-peak at mass 1133.6 amu.

In typical static SIMS experiments 1 $\mu$l of sample solution (concentration about $10^{-3}$ mol/l) is deposited on an etched silver surface and dried. After insertion of the silver target into the combined EI/SIMS cell via the solid sample inlet, the target is bombarded by $Cs^+$ for 0.02 - 10 msec with an intensity of 1-10 nanoAmp. Nearly all ions created by static SIMS are trapped at an ambient pressure of 1 - 5 x $10^{-9}$ mbar.

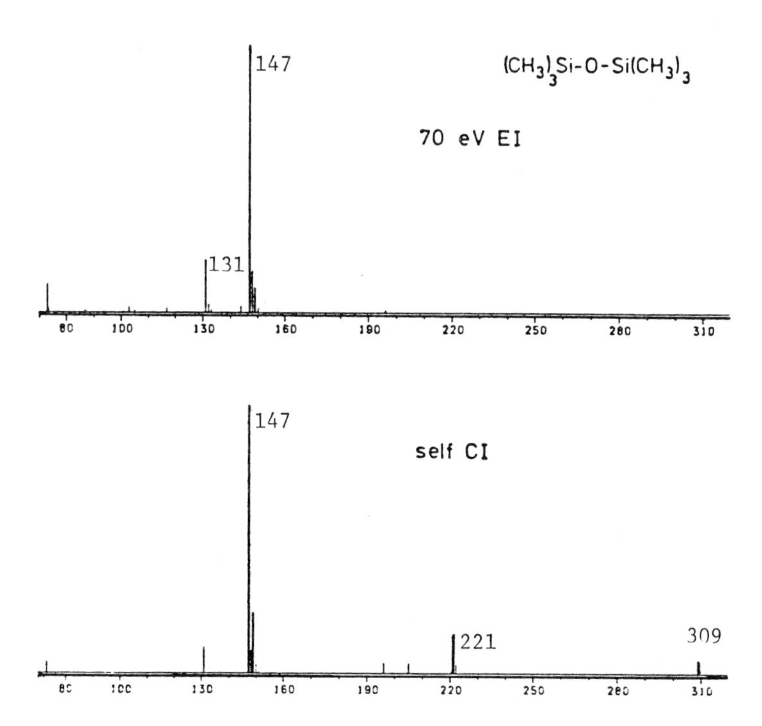


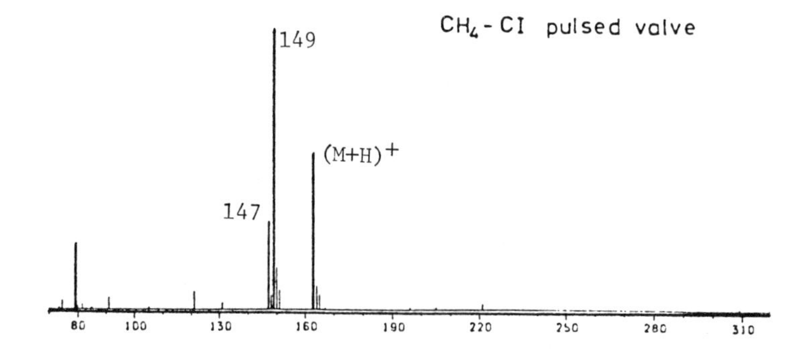


Figure 3a. 70 eV EI spectrum, self-CI, and pulsed-valve $CH_4$ CI positive ion spectrum of hexamethyldisiloxane.

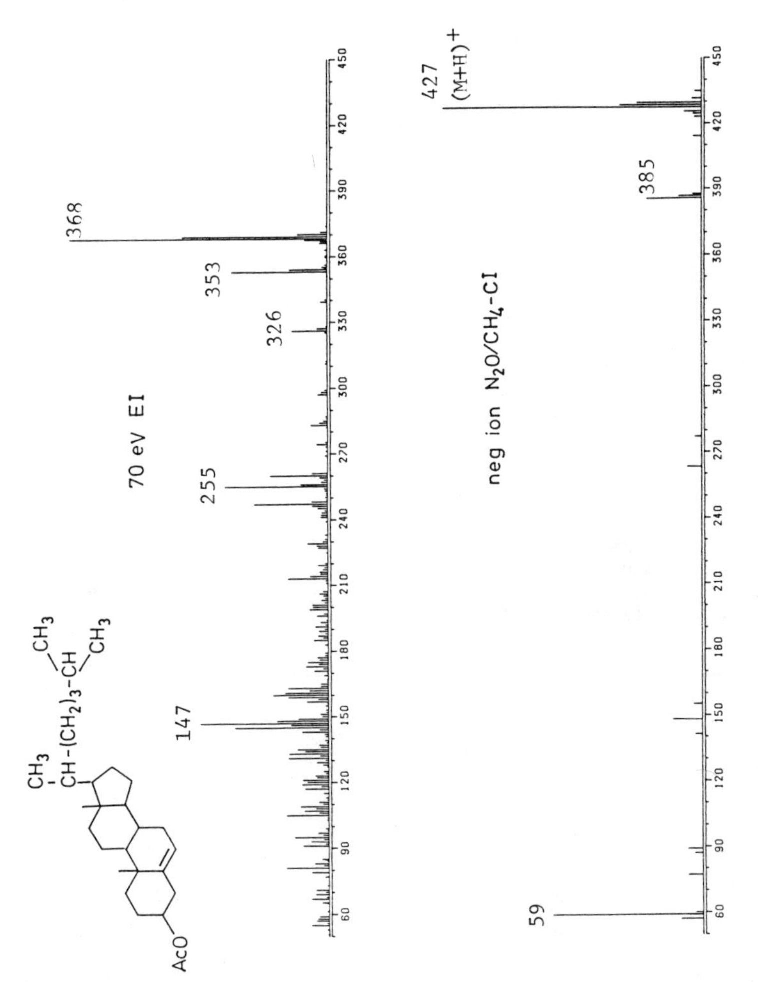


Figure 3b. 70 eV EI and pulsed-valve negative ion $N_2O/CH_4$-CI FTMS spectrum of cholesteryl acetate.

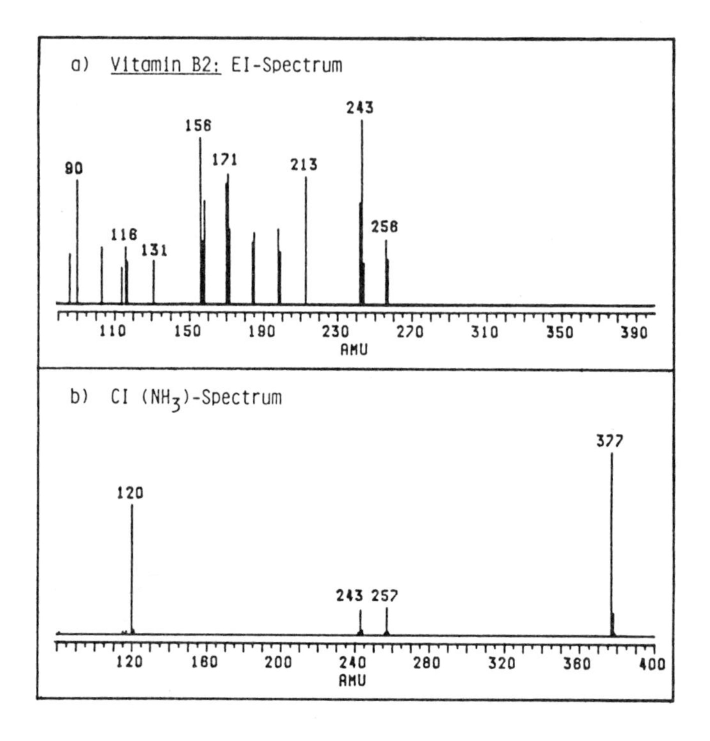


Figure 4. Low-pressure EI and DCI FTMS spectra of Riboflavin (Vitamin B2). Direct probe inlet, about 210°C.

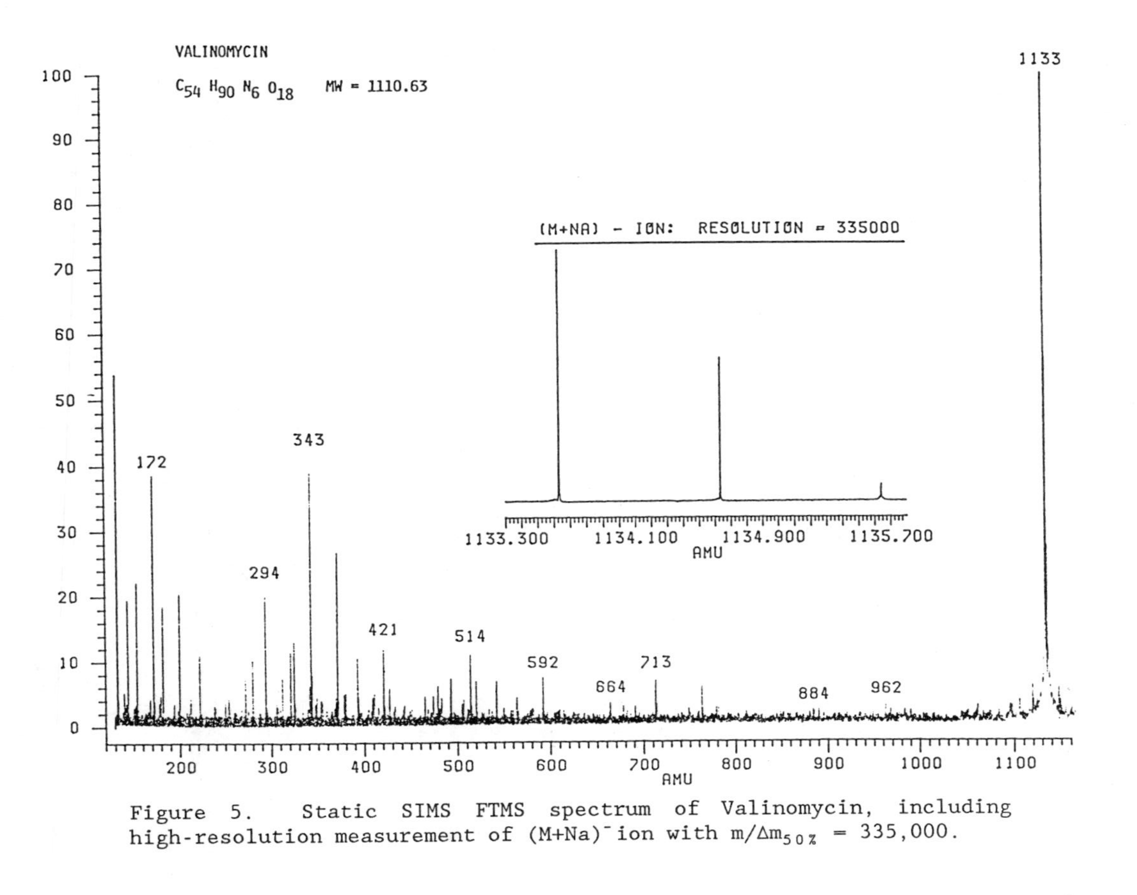


Figure 5. Static SIMS FTMS spectrum of Valinomycin, including high-resolution measurement of $(M+Na)^-$ ion with $m/\Delta m_{50\%}$ = 335,000.

Magnet Requirements for FTMS

In this section the importance of both the magnetic field strength B and of the magnetic field homogeneity will be discussed. Even though analytical FTMS experiments can be done in resistive, low-field magnets, we shall restrict our discussion to superconducting magnets, which permit superior mass resolving powers and absolute mass accuracies.

Using a horizontal 4.7 Tesla cryoshimmed supercon magnet with a bore diameter of 15 cm, we have obtained two-parameter mass calibrations with absolute mass accuracies of better than 1.5 ppm without, and better than 0.4 ppm with an internal calibrant over a mass range from 18 to 502 amu (10). Absolute accuracies with an internal calibrant are slightly better because the calibrant and the unknown compound are measured under identical physical conditions, in particular with the exact same number of ions in the cell, resulting in identical space-charge shifts (2).

Due to the extreme stability of supercon magnets and the associated electronics, 2-3 ppm accuracies can be reproduced in the absence of reference compounds for many days.

Field Strength. All other conditions being equal, the mass resolving power increases linearly with field strength according to (11)

$$m/\Delta m = qB\tau/m \quad (2),$$

where $\tau$ is the effective signal-decay time. If the linewidth is dominated by pressure broadening, as is typically the case for pressures above $10^{-9}$ mbar, particularly for higher masses over M = 100, then the resolution indeed improves linearly with B. In the collisional-broadening regime it is possible either to reduce the pressure or to increase the field strength in order to improve resolution.

The signal-to-noise ratio, S/N, in FTMS is given by (12)

$$S/N = \frac{Nq^2 \rho B \sqrt{R}}{2md_x \sqrt{2kT\Delta f}} \quad (3),$$

if amplifier noise is neglected. Here N is the total number of ions, $\rho$ is the cyclotron radius, R is the resistance, T is temperature, and $\Delta f$ is the detection bandwidth. Clearly, higher field strength increases the FTMS sensitivity linearly. Nevertheless, the increase of S/N with B is not as dramatic as in NMR, where sensitivity increases with the seven-fourth power of the field strength (13).

If one is interested in high-mass FTMS it is advantageous to work at the highest possible field strength, because the higher the cyclotron frequency of a given mass, the less noise is introduced during signal preamplification. Moreover, the gain of most rf power amplifiers used for acceleration, as well as the gain of the signal amplification circuitry, tend to fall off for lower frequencies, i.e. higher masses.

Finally, Equation 1 exhibits another important advantage of high fields in FTMS. Both the maximum trapping time and the maximum number of collisions (and gas-phase reactions) increase quadratically with B. Consequently increased magnetic field strength offers experimental access to larger ion clusters (Figure 2).

Homogeneity. Residual spatial magnetic field inhomogeneities limit the maximum mass resolving power to a pressure-independent upper limit, despite the fact that certain inhomogeneities are averaged by the ion motion in the cell (14). The experimental and theoretical limit of $m/\Delta m$ in our 4.7 Tesla cryoshimmed supercon magnet is approximately $10^8$, as shown in Figure 6, where positive methane ions were measured with a mass resolving power of $2 \times 10^8$. In superconducting magnets without cryoshims this limit is likely to be one order of magnitude lower, and in electromagnets the upper resolution limit is about $10^5$. Theoretical calculations (14) show that the maximum mass resolving power can increase by a factor between 20 and 100 to $10^9$ - $10^{10}$, if room-temperature shims are used in addition to the standard cryoshims.

## Software Requirements for FTMS

Starting from the general philosophy that an FTICR mass spectrometer is a research grade analytical instrument, we have chosen to develop a software system which offers versatility and direct control over almost all experimental parameters.

Phasing. In order to take full advantage of FTICR's high-resolution capabilities, automatic and manual phasing have been implemented from the beginning. Consequently, in the high-resolution mode absorption spectra which require phasing can be obtained instead of magnitude spectra. Marshall (15) has pointed out that absorption spectra are narrower than magnitude spectra by a factor of $\sqrt{3}$. An example of automatic zeroth-order, and interactive first- and second-order phasing of an absorption mode spectrum is given in Figure 7, and a typical absorption mode high-resolution spectrum at mass M = 131 is shown in Figure 8. The phasing problem in broadband FTMS has not yet been solved, and usually only magnitude spectra are displayed.

Rapid Scan/Correlation ICR. In order to provide additional experimental flexibility, the CMS 47 can be operated in the rapid scan/cross correlation mode (16,17) without any hardware modifications. In this mode the receiver plates of the ICR cell are incorporated in an rf capacitance bridge circuit which detects energy absorption of the ions during a low-power rapid frequency scan using sweep rates of approximately 1 MHz/s. Rapid scan/correlation ICR can be utilized advantageously when very large mass ranges are to be detected, and particularly when large mass ranges extend into the low mass/high frequency region below mass M = 20. An example of a cross correlation spectrum is given in Figure 9. With standard FT-NMR phasing techniques both absorption and magnitude spectra can be obtained (16). Rapid scan/cross correlation ICR has the additional advantage that intensity ratios are reproduced more accurately than in FTICR.

Versatile Pulse Sequence. One of the great strengths of FTMS is the flexibility to selectively accelerate, activate, and eject ions in any combination and any sequence without hardware modifications. This versatility makes FTMS the method of choice for MS/MS and hence for establishing pathways and rate constants for gas-phase ion-molecule reactions, and to correlate this data with structural information. Recently up to $(MS)^5$ has been demonstrated (18).

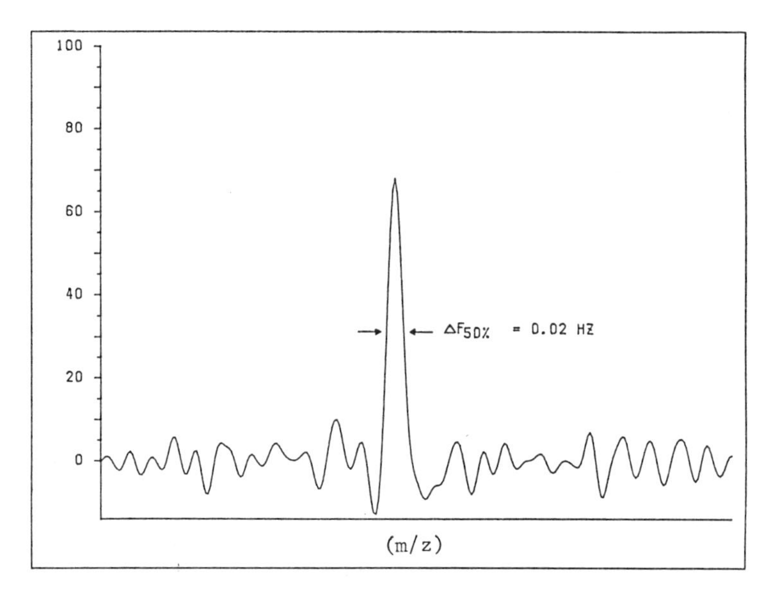


Figure 6. Ultra-high resolution FTMS spectrum of positive methane ions with $m/\Delta m_{50\%} = 2 \times 10^8$, $\Delta f_{50\%} = 0.02$ Hz.

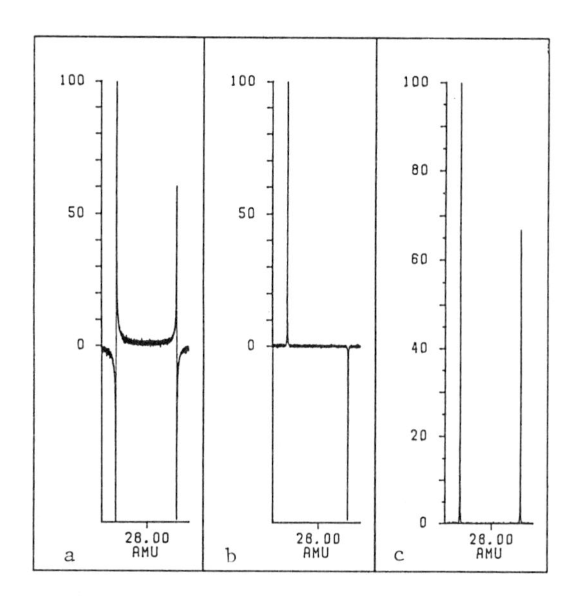


Figure 7. High-resolution absorption mode peaks of $N_2^+$ and $CO^+$: a) unphased, b) after automatic zeroth-order phasing, c) after interactive first- and second-order phasing.

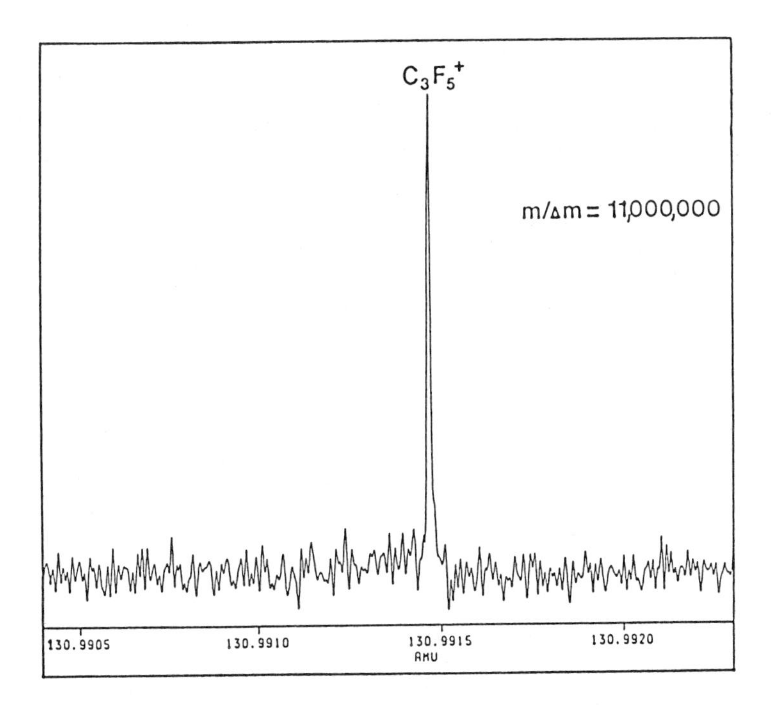


Figure 8. High-resolution absorption mode spectrum of $C_3F_5$+ at mass 131 amu. Resolution $m/\Delta m_{50\%}$ = 11,000,000.

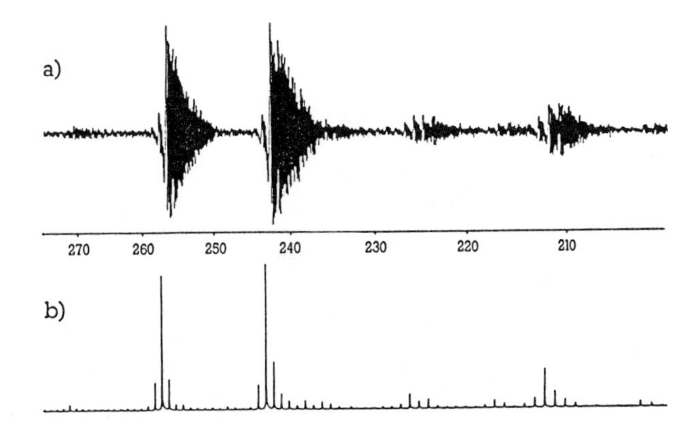


Figure 9. a) Rapid scan spectrum b) cross correlation spectrum.

To exploit the advantages of FTMS fully we have implemented several predefined ejection, activation and acceleration options, like single shots, covers, sweeps over some mass window, sweeps around some unaccelerated mass window, tickling belts with phase inversion, parent ion selection chirps, activation shots for daughter ion production, etc.

Using automation software routines for the ASPECT 3000 pulse generator, we can utilize up to nine ejection/activation pulses, each combining up to four options. Mass (or frequency) information is taken from several variable user-defined lists, such that each option in each event may be used on a large number of masses. The variable mass and delay lists can be defined either via alphanumeric keyboard input, or they can be defined interactively from the spectrum via cursor. Each acceleration/ejection event can contain up to 4095 different steps. All events can be separated by fixed or incrementable delays.

*Computer Interpretation, Timesharing and Throughput*. For routine applications and high throughput, timesharing is essential. On the CMS 47 three so-called jobs can run in timesharing mode. For instance, job 1 may be acquiring data, job 2 may plot previously taken data, while the user is defining the parameters of the next experiment in job 3 in a menu-driven fashion.

Finally, for routine applications, our software provides a database management system called BASIS for storage and manipulation of chemical information. BASIS can access generally available spectral libraries from three different spectroscopic techniques (MS, H-NMR and F$^{1}$3C-NMR, IR), and permits the creation of new libraries. For structure elucidation and substructure search of unknown compounds, library search algorithms allow the retrieval of identical and structurally similar spectra.

### External Ionization

Typical pressures in the analyzer region of FTICR mass spectrometers are about three orders of magnitude lower than in most other types of mass spectrometers. For instance in our single cell CMS 47 we can routinely obtain pressures of $3x10^{-10}$ mbar using a 330 l/sec Balzers turbomolecular pump. It is very difficult to couple existing, high gas-load GC-MS, LC-MS, liquid SIMS, or FAB ion sources directly into a single-cell configuration. Other ionization techniques like static SIMS and Laser Desorption (LD) can be operated with much improved throughput in a medium-pressure source.

It is therefore evident that a differentially-pumped ion source, operating at $10^{-5}$ to $10^{-6}$ mbar, is a desirable feature of an FTICR mass spectrometer, particularly for routine applications. Three approaches are presently followed to implement differentially-pumped ion sources: the dual cell which is described elsewhere in this book, and external ionization with (19) and without (20) tandem quadrupoles.

External ionization has the advantage of completely removing the ionization region from the narrow magnet bore while retaining all the features which make FTMS such an attractive technique, like high resolution, accurate mass determinations, wide mass range, long trapping times, selective ejection/activation/acceleration etc. Our external ion source, which uses simple accelerating and focusing elements, greatly improves access to the ionizer, thereby facilitating the coupling of various inlet systems, without the need

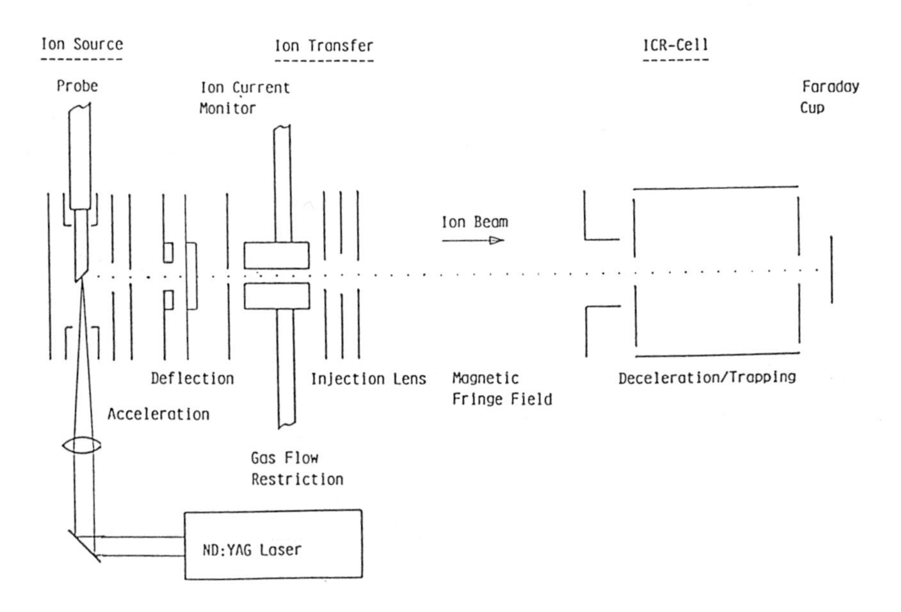


Figure 10. Schematics of ion transfer mechanism from differentially pumped external ion source with LD.

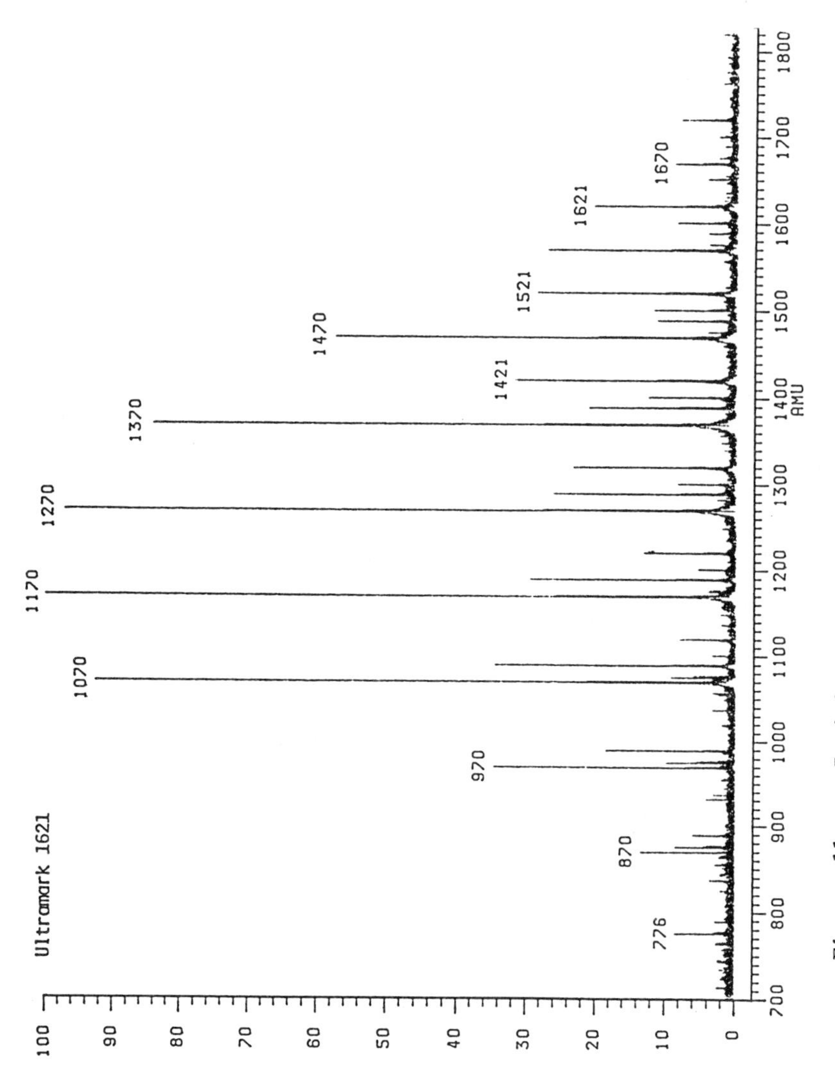


Figure 11a. Positive ions of Ultramark 1621. Electron impact ionization with direct insertion probe in external ion source.

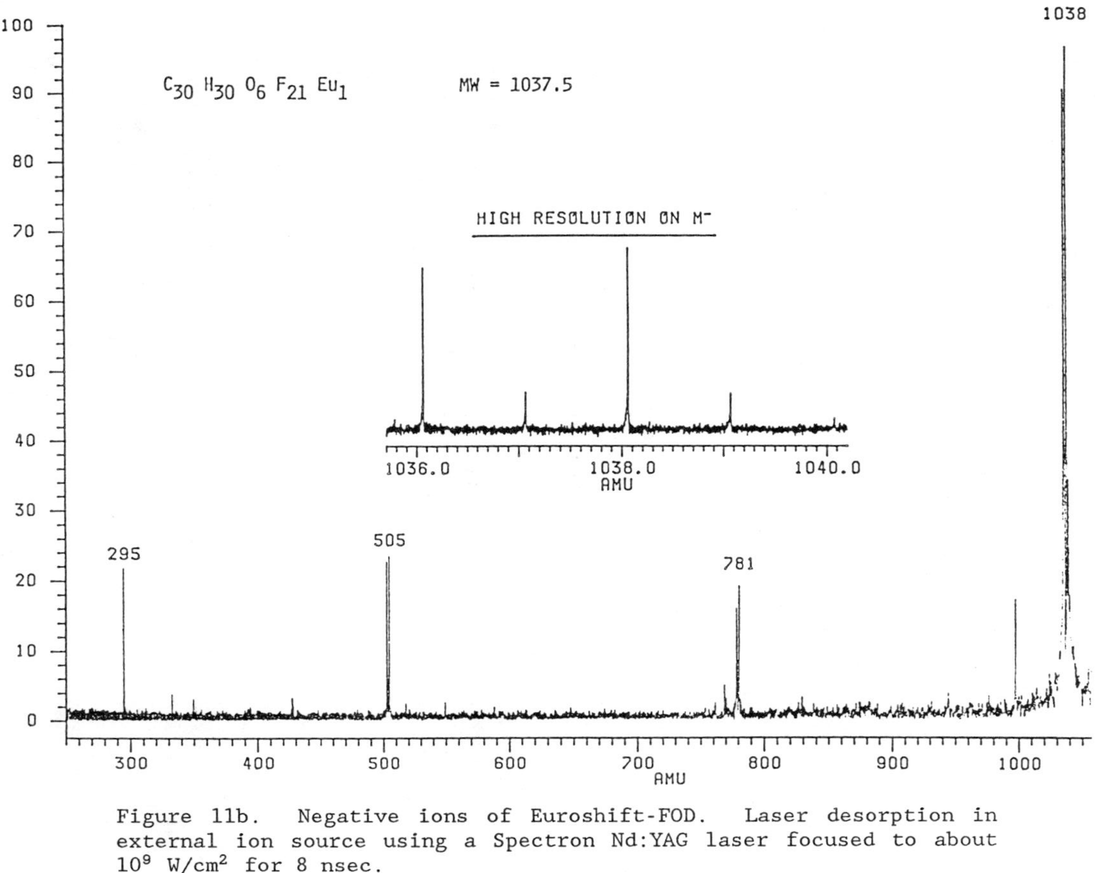


Figure 11b. Negative ions of Euroshift-FOD. Laser desorption in external ion source using a Spectron Nd:YAG laser focused to about $10^9$ W/cm² for 8 nsec.

for expensive and experimentally more complex rf quadrupoles. It does not provide for mass preseparation outside the magnet, but mass selection can be obtained with FTMS double resonance techniques. Typical ion current transmission efficiencies are of the order of 30%. The ion transfer efficiency from source to analyzer is mass-independent (20). For special analytical applications, time-of-flight effects during ion transmission can be utilized to separate certain mass windows by setting appropriate time intervals between EI or LD ionization and the cell trapping pulse (21).

Figure 10 shows a schematic drawing of the ion transfer mechanism in the CMS 47 with external ionization. Figure 11a presents the mass spectrum of Ultramark 1621 using electron impact in the external source, and Figure 11b depicts a broadband spectrum and the high-resolution window of the negatively charged molecular ion of Euroshift-FOD after laser desorption in the external ion source. Moreover, the external ion source can be employed for medium-pressure CI and SCI as an alternative to the low-pressure and pulsed-valve CI techniques described above.

## Anticipated Developments in FTMS Instrumentation

With the successful implementation of differentially-pumped external ion sources, FTMS is rapidly becoming a routine mass spectrometric technique. Medium-pressure interfaces for the coupling of GC, LC, FAB, and liquid SIMS into the external ionizer are currently under development, and should become available in the near future.

Recently, new 2D-methods for the analysis of complex mixtures have been developed for time-of-flight mass spectrometry (22), which could also be utilized in external ionization FTMS. Specifically, the combination of IR-laser desorption of nonvolatile neutrals, followed by adiabatic cooling to 2°K in a supersonic jet, and subsequent compound-selective Resonance-Enhanced Multiphoton Ionization (REMPI) could increase the role of FTMS in the analysis of biological mixtures. The coupling of supersonic jets to the external ion source would also be of interest in ion- and neutral cluster experiments.

Finally, it is conceivable that ultra-high resolution FTICR will find additional applications in particle, nuclear, and atomic and molecular physics. We believe that some of the instrumental prerequisites for these applications will include a) working at lower pressures below $10^{-11}$ mbar, b) better magnetic field homogeneity using additional room-temperature shims and employing improved shimming techniques (14), and c) ultra low-noise preamplifiers capable of detecting single or very few ions.

## Acknowledgments

FHL wishes to thank Professor William Klemperer at Harvard University for his encouragement and guidance. The authors thank Robert A. Forbes for his suggestions and for reading the manuscript. Finally the authors are grateful to Michelle V. Buchanan for her tremendous efforts in organizing this FTMS book.

## Literature Cited

1. Wanczek, K.P. Int.J. Mass Spectrom. Ion Processes 1984, 60, 11-60.

2. Jeffries, J.B.; Barlow, S.E.; Dunn, G.H. Int.J. Mass Spectrom. Ion Processes 1983, 54, 169.
3. Francl, T.J.; Fukuda, E.K.; McIver, R.T. Int.J. Mass Spectrom. Ion Physics 1983, 50, 151.
4. Dunbar, R.C. Int.J. Mass Spectrom. Ion Processes 1984, 56, 1.
5. McIver, R.T. 30th An. Conf. Mass Spectrom. All. Top. A.S.M.S., Honolulu 1982.
6. Freiser, B.S. 30th An. Conf. Mass Spectrom. All. Top. A.S.M.S., Honolulu 1982.
7. Zwinselman, J.J.; Allemann, M.; Kellerhals, Hp. Spectrospin ICR Application Note 1984, IV.
8. Cody, R.B.; Freiser, B.S. Anal. Chem. 1982, 54, 1431.
9. Grossmann, P.; Allemann, M.; Kellerhals, Hp. Spectrospin ICR Application Note 1986, VI.
10. Klass, G.; Allemann, A.; Bischofberger, P.; Kellerhals, Hp. Spectrospin ICR Application Note 1983, II.
11. Marshall, A.G.; Comisarow, M.B.; Parisod, G. J. Chem. Phys. 1979, 71 (11), 4434.
12. Comisarow, M.B. J. Chem. Phys. 1978, 69 (9), 4097.
13. Hoult,D.I.; Richards,R.E. J. of Magn. Reson. 1976, 24, 71.
14. Laukien,F.H. Int.J. Mass Spectrom. Ion Processes 1986, 73, 81.
15. Marshall,A.G. In Fourier, Hadamard and Hilbert Transforms in Chemistry; Marshall,A.G.,Ed.; Plenum Press: New York, 1982; Chapter 1.
16. Parisod,G.; Gaumann,T. Chimia 1980, 34, 271.
17. McIver,R.T.; Hunter,R.L.; Ledford,E.B.; Locke,M.J.; Francl,T.J. Int. J. Mass Spectrom. Ion Phys. 1981, 39, 65.
18. Forbes,R.A.; Laukien,F.H.; Wronka,J. submitted to Int.J. Mass Spectrom. Ion Processes.
19. Hunt,D.F.; Shabanowitz,J.; Yates,J.R.; McIver,R.T.; Hunter,R.L.; Syka,J.E.P.; Amy,J. Anal. Chem. 1985, 57, 2728.
20. Kofel,P.; Allemann,M.; Kellerhals,Hp. Int.J. Mass Spectrom. Ion Processes 1985, 65, 97.
21. Kofel,P.; Allemann,M.; Kellerhals,Hp. Int.J. Mass Spectrom. Ion Processes 1986, 72, 53.
22. Frey,R. 34th An. Conf. Mass Spectrom. All. Top. A.S.M.S., Cincinnati, 1986.


RECEIVED June 15, 1987

Chapter 6

# Fourier Transform Mass Spectrometry of Large ($m/z > 5,000$) Biomolecules

**Curtiss D. Hanson[1], Mauro E. Castro[1], David H. Russell[1], Donald F. Hunt[2], and Jeffrey Shabanowitz[2]**

**[1]Department of Chemistry, Texas A&M University, College Station, TX 77843**
**[2]Department of Chemistry, University of Virginia, Charlottesville, VA 22901**

Recent experimental results demonstrate the high mass (>10,000 amu) capabilities of Fourier transform mass spectrometry, however the data reveal non-theoretical limits in the resolution at high mass. These advances can be attributed to the development of methods for coupling high pressure ion sources to the ultra-high vacuum FT-ICR analyser. Specifically, external ion sources permit the utilization of liquid matrix secondary ionization mass spectrometry for the desorption of large involatile, thermally labile biomolecules with ion detection by FT-ICR. Limitations in the mass resolution arise from the inability to effectively trap the ions and produce a coherent packet of ions for detection. The lack of spatial and phase coherence of the injected ions leads to the loss of the frequency domain signal. According to our model, a principle factor contributing to the lack of spatial and phase coherence is field inhomogeneities coupled with the kinetic energies of the ions along the Z-axis.

The development of an ion excitation scheme for ion cyclotron resonance (ICR) compatible with Fourier transform data analysis methods has greatly increased the analytical utility of the method.[1,2] Although the potential utility of Fourier transform mass spectrometry (FTMS) for the analysis of large biomolecules was recognized early,[3] the development of suitable ionization methods and experimental hardware for biomolecule FTMS proved to be a rather difficult task.[4,5] One consideration in designing a system for analysis of biomolecules is the high vacuum requirements of ion detection by FT-ICR methods.[1,6] The requirements for maintaining high vacuum ($10^{-8}$ torr or less) for ion trapping and high resolution mass measurements has led to the adaption of ionization methods for FTMS such as laser desorption[4] and $Cs^+$ ion SIMS.[5] Although these ionization methods are quite useful for the analysis of some biomolecules, the success with molecules larger than 2,500 daltons has been limited.

0097–6156/87/0359–0100$06.00/0

The introduction of the tandem quadrupole-Fourier transform mass spectrometer[7,8], the dual cell analyzer,[9] and the external ion source[10] opens new possibilities for analyzing biomolecules by FTMS. The first real success with biomolecules was sample ionization by liquid matrix SIMS and introduction of the sample ions to the ion cell by the tandem quadrupole ion injection system.[7] Although the transmission efficiency (fraction of ions trapped in ion cell relative to the total secondary ion yield) of the device has not been fully characterized, especially for large molecules, it is clear that the sample detection levels are comparable to the most sensitive magnetic sector instruments and time-of-flight mass analyzers.[8,11] An additional advantage of the tandem quadrupole-FTMS system (Q-FTMS) has been recently demonstrated by the photodissociation of large peptides.[11]

The work performed on the Q-FTMS reveals a fundamental problem with the method, viz., the inability to preform high resolution mass measurement at high mass.[11] Although impressive mass resolution data has been reported for ions of 1500-2000 daltons, the limited duration of the frequency domain signal for ions above m/z 2500 is insufficient for high resolution measurements.[11] Similar limitations apply to solid-state SIMS experiments performed in a Nicolet FTMS-1000 system equipped with a single section ion cell.[12] In an attempt to understand this problem, we will consider the physics of trapping ions in the ion cell and the effect of the initial velocity of the ions and field inhomogeneities on the ion trapping.

## Detection of High Mass Ions using FTMS

A major thrust of recent work on FTMS of large biomolecules has dealt with questions concerning the lifetime of molecular ions formed by high-energy particle bombardment. Chait and Field have reported that a large fraction of the molecular ions of chlorophyll A formed by $^{252}$Cf fission fragment ionization decomposes with lifetimes of less than a few microseconds.[13] A later study on the $[M+H]^+$ ion of insulin showed that extensive dissociation occurs on both the nanosecond and microsecond time scale.[14] These results raise questions concerning the utility of FTMS for the analysis of large biomolecules. For example, to acquire a low resolution FT mass spectrum from m/z 100 to m/z 10,000 requires data acquisition times of 300-500 ms. On the basis of Chait and Field's study it is reasonable to suspect that molecular ions of large molecules may not survive the long times required for FT detection. The problem is even more severe when considering the requirements for acquiring high mass resolution data at high mass, e.g., conditions that require data acquisition times of several seconds to even tens of seconds.

Earlier studies on the detection of $Cs(CsI)_n^+$ cluster ions were directed at addressing the general problem discussed above.[15] Although there is ample sensitivity to detect high mass (>m/z 10,000) $Cs(CsI)_n^+$ cluster ions in the low and high mass resolution modes, there are questions concerning how well the behavior of such cluster ions models the behavior of large organic molecules.

The analysis of biomolecules by FTMS methods at Texas A&M University has been limited to $Cs^+$ desorption ionization from a solid-state matrix in order to maintain the high vacuum requirements. Due to the low ion yields for $[M+H]^+$ type ions by keV energy particle bombardment from the solid-state, the number of successful analyses with molecules larger than 2000-2500 amu are limited.[12] Owing to the differential pumping arrangement of the ion source and the analyzer, sample ionization on the tandem quadrupole-FTMS instrument at the University of Virginia can be performed by desorption ionization from a liquid matrix, i.e., keV energy $Cs^+$ ion particle bombardment of a liquid matrix. Early studies with this instrument demonstrate both the sensitivity and mass resolution of the method.[8,11]

Recent modifications to the Q-FTMS instrument have greatly increased the signal-to-noise ratio and resolution for high mass samples.[11] The modifications involved milling out the ion source block to increase the gas flow in the vicinity of the sample probe. Aberth has discussed the importance of gas pressure on the performance of particle bombardment ion sources.[16] The second modification involved pulsing of the trapping plate voltage of the ion cell from 1 to 10 volts during the ionization step. These two relatively simple modifications have increased the sensitivity of the instrument by roughly two orders of magnitude and extended the working mass range for organic samples beyond mass 10,000.[11]

The data for neurotensin (a tridecapeptide) shown in Figure 1 illustrate the available sensitivity and mass resolution. This spectrum was obtained from a 100 pmol sample dissolved in thioglycerol/glycerol matrix and by using a primary $Cs^+$ ion beam irradiation time of 4 ms. The secondary ions formed on particle bombardment were injected into the ion cell by the rf-only quadrupole rods operated with a low-mass cut-off of ca. m/z 400. As noted above, the trapping voltage was set to zero during ionization and then pulsed positive (to 3 volts for this sample) following the ionization step.[11] The purpose of pulsing the trapping voltage positive following ionization is to restrict the motion of the ions in Z-direction (vide infra).

In the broadband (low mass resolution) mode the only ion detected was the $[M+H]^+$ ion, and in the narrowband mode a mass resolution of ca. 90,000 (resolution specified at FWHM) was obtained. Note also that the signal-to-noise ratio in the broadband mode is > 100:1 for a sample size of 100 pmol, and in the narrowband (high resolution) mode the signal-to-noise ratio is greater than 500:1. Owing to the pulsed ionization used for FTMS, the total ionization time required to produce this spectrum was approximately 80 ms. That is, the $Cs^+$ ion beam was turned on for 4 ms to produce each mass spectrum and 20 individual ionization events were signal-averaged.

Tandem quadrupole-FTMS data for two 2000-6000 mass range peptides are shown in Figure 2. Melittin (26 residues) and glucagon (29 residues) were analyzed in earlier work, but even with long sample ionization times (5 s) the signal-to-noise in the region of the $[M+H]^+$ was low (<10).[8] In the spectrum shown in Figure 2a total sample ionization time of 80 ms was employed, e.g., 4 ms bombardment with $Cs^+$ and signal-averaging of 20 broadband spectra.

Three more recent results clearly establish the high mass range capabilities of the FTMS experiment. Spectra for bovine insulin (Figure 3), a mixture of bovine insulin and porcine insulin (Figure 4), and horse cytochrome-C (Figure 5) were obtained on sample sizes of 100 pmol, 100 pmol (equimolar), and 250 pmol, respectively, and with sample ionization times of either 80 ms and 800 ms (cytochrome-C). For the insulin samples abundant doubly charged ions were observed, and in both cases the B-chain fragment ions were detected. In the case of horse cytochrome-C a doubly charged ion was not observed with sufficient signal-to-noise for assignment.

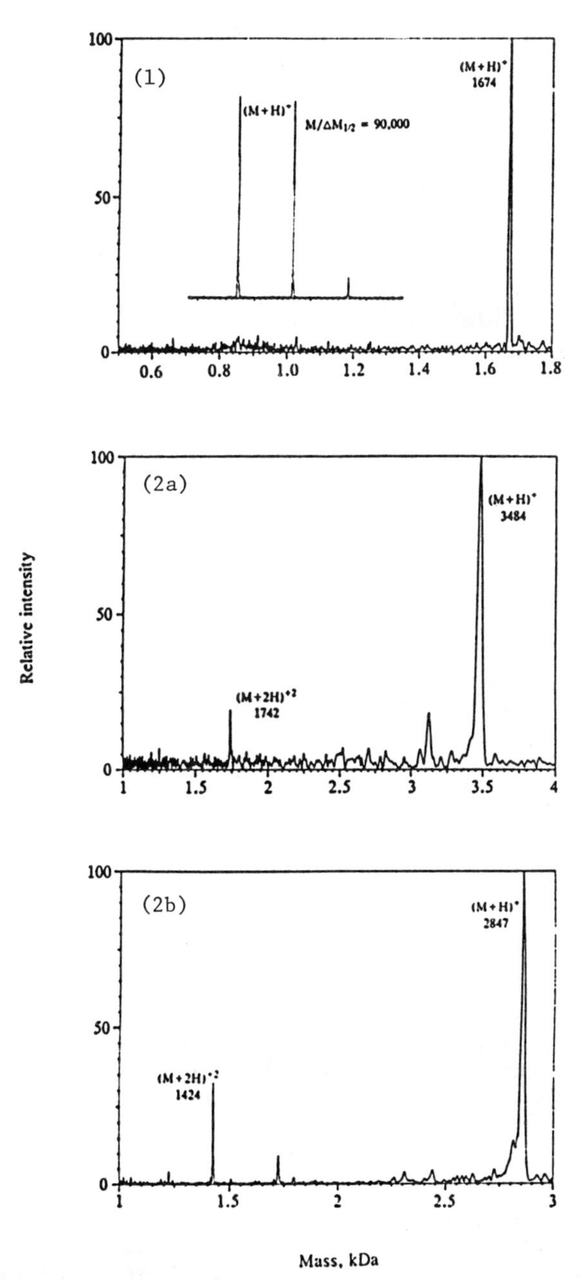


Figures 1-2. Liquid secondary-ion mass spectra. (1) Neurotensin (100 pmol); $M_r$=1672.9. (2) a) Glucagon (100 pmol); $M_r$=3482.8 b) Melittin (100 pmol); $M_r$=2846.5.
(Reproduced with permission from Ref. 11. Copyright 1987 National Academy of Sciences.)

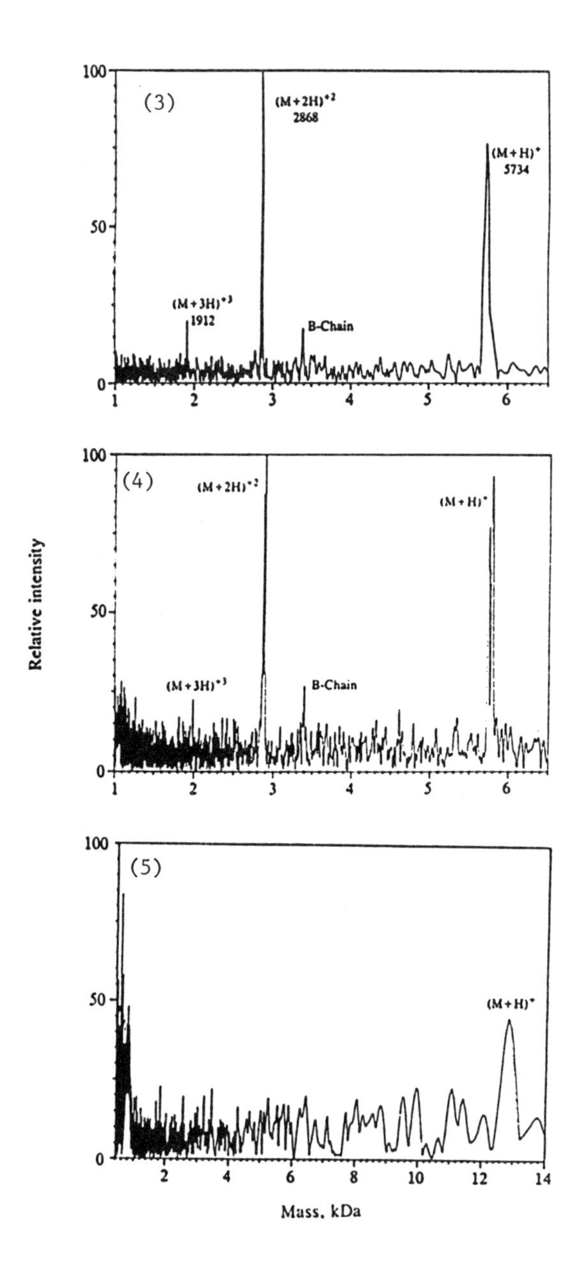


Figures 3-5. Liquid secondary-ion mass spectra. (3) Bovine insulin (100 pmol); $M_r$=5733.5. (4) Bovine ($M_r$=5733.5) and porcine ($M_r$=5777.6) insulins (200 pmols each). (5) Horse Cytochrome c ($M_r$=12,384).
(Reproduced with permission from Ref. 11. Copyright 1987 National Acedemy of Sciences.)

Although the data for the insulin and cytochrome-C samples clearly illustrate the mass range capabilities of FTMS, two significant problems with the experiment must be addressed further. These problems are (i) the inability to monitor the image current for large peptides for sufficiently long times to permit high resolution mass measurements, and (ii) the difficulties in accumulating large organic ions in the ion cell from multiple ionization events. That is, in order to increase the sensitivity of biomolecule FTMS and to minimize thermal degradation of the sample by the ionization radiation, one would like to accumulate ions from several short (microsecond to nanosecond) ionization pulses and then perform a single detection step. The ion accumulation experiment could also prove beneficial for laser-photodissociation (as opposed to collision-induced dissociation) for structural characterization of large molecules. Since the present work demonstrates the stability of large organic ions on the millisecond time scale (detection times for the broadband mode, e.g., acquisition times for the data in Figures 2-5 range from 100-500 ms), it seems unlikely that the ion population is depleted by slow dissociation reactions on this time scale. Thus, the nanosecond and microsecond dissociation reactions studied by Chait and Field[13-14] should not be a complicating factor for either mass resolution or ion accumulation.

If the inability to perform high mass resolution and the ion accumulation experiment at high mass is not related to the dissociation dynamics of larger organic ions then the problem must lie in either ion trapping or ion detection. To address this problem we will examine the physics of ion motion in a trapped ion cell in a strong magnetic field.

## Ion Production, Trapping, and Detection in an ICR Ion Cell

The most commonly used ionization mode for ICR is electron impact, and the basic principles of ICR and FTMS have been developed and tested by using this ion production method. Ions produced in this manner have near-thermal translational energies, and the ions cyclotron at the natural cyclotron frequency ($w = qB/m$) in the X-Y plane. The motion of the ions in the Z-direction is constrained by the potentials applied to the two trap plates positioned in the X-Y plane which define the Z-axis limits of the ion cell. The initial dimensions of the ion population in the X-Y plane and Z direction are defined by the dimensions of the electron beam (fixed by the diameter of the electron beam in the X-Y plane, ca. 1 mm) and ion cell (distribution along the Z-axis is determined by the length of the ion cell) (Figure 6). After the electron beam is turned off the ions will redistribute in the ion cell. The largest distribution of the ion population will be along the Z-axis. Under collisionless conditions the ions will oscillate in the Z-direction. The amplitude of the motion in the Z-direction will be determined by the initial conditions of the neutral prior to ionization and the point at which the ion is formed. That is, ions formed near the Z-direction trapping plate will have a greater amplitude of longitudinal oscillation.

Diffusion of the ion population in the X-Y plane will depend upon the ion density (space-charge effects) and the strength of the applied magnetic field. Under conditions of high vacuum and in the absence of space-charge effects, diffusion of the ion population in the X-Y plane does not impose any serious problems for ion detection. That is, the radio-frequency signal

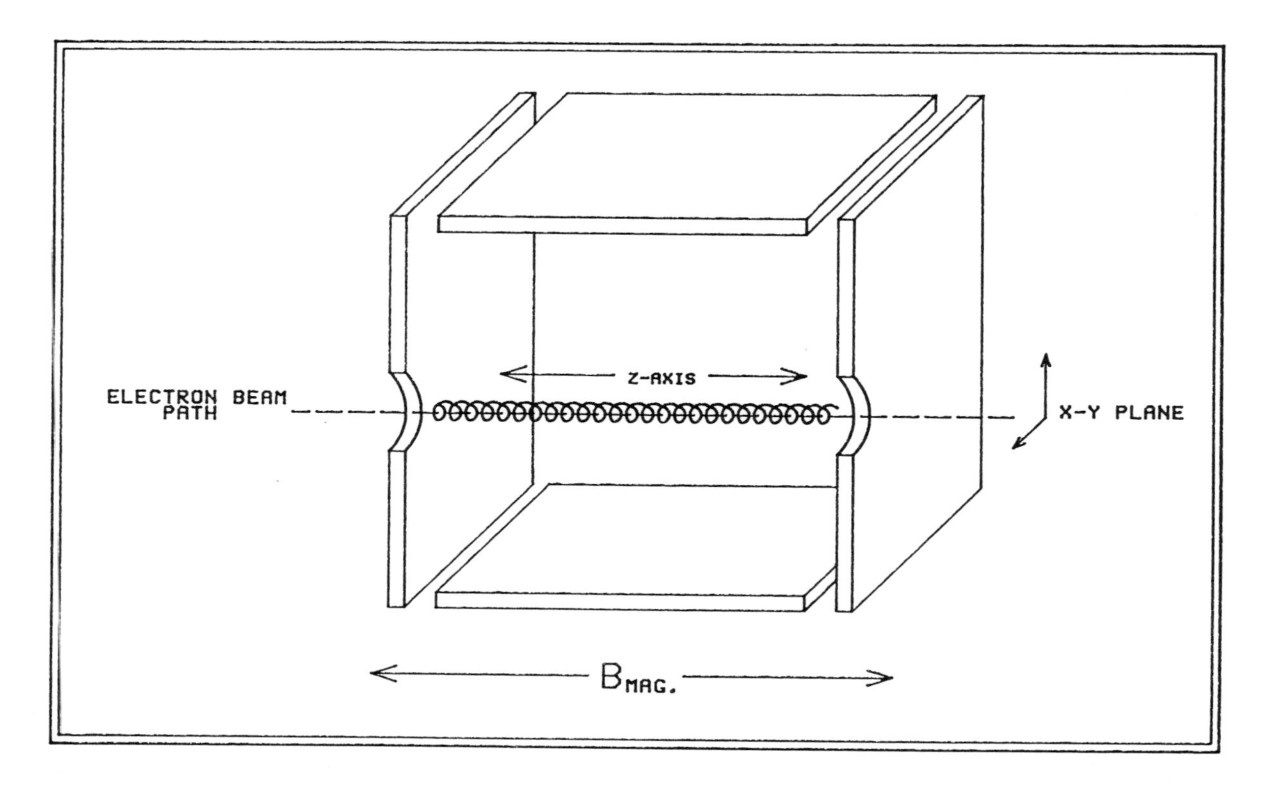


Figure 6. The initial dimensions of the ion population as defined by the diameter of the electron beam and the length of the cell.

applied to excite the ion population for detection accelerates the entire ion packet to a larger radius orbit, the ion population remains bunched together, i.e., a coherent packet of ions (Figure 7). It is this packet of coherently cycloiding ions which induces the image current in the receive plates and thereby produces the frequency domain signal.

If a sufficiently long delay is introduced between ionization and ion detection, diffusion of the ion packet in the X-Y plane can occur. Excitation of such an ensemble of ions will result in a random spatial distribution but a coherent phase relationship (Figure 8). This coherent phase relationship will result in the generation of a frequency domain signal. Conversely, if the ensemble of ions is not phase coherent due to the combined motion in the X-Y plane, Z-direction, and/or problems associated with the excitation process, loss of signal will occur. In the presence of homogeneous electrostatic and magnetic fields with ions of near zero kinetic energies along the Z-axis, the rate of ion diffusion in the X-Y plane will be quite low. However the rate of diffusion and subsequent loss of coherence can be increased by the influence of field inhomogeneities coupled with a significant kinetic energy spread of ions along the Z-axis. This loss of signal becomes crucial when the quantity of ions of interest produced during the ionization event is already small prior to the subsequent spatial and phase randomization.

It is important to keep in mind that electrostatic and magnetic inhomogeneities can have three effects on the FTMS experiment. First, field inhomogeneities can influence ion trapping, i.e., the ability to maintain a stable density of ions in the ion cell and to accumulate ions from successive ionization steps. Second, field inhomogeneities can contribute to signal broadening and loss of mass resolution. The second point has been examined both theoretically and experimentally and the effects of magnetic field inhomogeneity are quite small, even negligible.[17-19] It should not be surprising that small magnetic field inhomogeneities have little effect on ion detection. The ions are detected following excitation by an applied radio-frequency field, thus the ions have appreciable translational energies (in the X-Y plane). On the other hand, ions which are trapped in the ion cell have low translational energies, in both the X-Y plane and the Z-direction, and even small field inhomogeneities will have an appreciable effect on the ion motion. Thirdly, if the ions must pass through a region of field inhomogeneity, as is the case for ion partitioning in a two-section ion cell and for ion injection from an external ion source, ion losses may occur due to scattering in the X-Y plane. In terms of ion trapping, an additional aspect to consider is the initial velocity of the ion (in the X-Y plane and the Z-direction) as it enters the ion trap, e.g., from an external ion source (influencing the Z-axis motion) or from a tandem mass analyzer such as a quadrupole mass filter (influencing the velocity in both the X-Y plane and the Z-direction). Specifically, detection of ions produced external to the ion source may not be adequately modelled by studies of ions formed by EI.

The basic physics of ion motion in electrostatic and magnetic fields are well understood.[20] An ion formed in a magnetic field will acquire a circular orbit in the X-Y plane, and the frequency (w) of this orbit is dependent upon the magnetic field strength (B), charge on the ion (q), and inversely dependent on the mass (m) of the particle (Equation 1).

$$w = qB/m \quad (1)$$

In the absence of an applied electric field, the ion will also move freely in the Z-direction. Thus, the overall motion of the ion is best described as a helix (Figure 9). The radius of the helix (i.e., the radius in the X-Y plane) depends on the velocity of the ion in the x and y directions, i.e., $v_{ox}$ and $v_{oy}$, respectively, and is given by Equation 2.

$$R = w^{-1}[(v^2_{ox} + v^2_{oy})^{1/2}] \quad (2)$$

$$\propto m/qB[(v^2_{ox} + v^2_{oy})^{1/2}]$$

The pitch (h) of the helix is determined only by the velocity of the ion in the Z-direction ($v_{oz}$) and the period (T) of the ion motion.

$$h = Tv_{oz} \propto mv_{oz}/qB \quad (3)$$

where T is given by Equation 4;

$$T \propto 1 / w \propto m/qB \quad (4)$$

The effect of imposing an electric field parallel to the magnetic field, i.e., the situation as it is in a typical ICR ion trap, is to produce an ion motion in which the pitch varies continuously along the Z-axis (Figure 10). Such a motion will have an $m^{1/2}$ dependence (velocity dependent); whether the pitch is lengthened or shortened will depend upon the applied trapping potential.

Having introduced the rudiments of ion motion in a typical FTMS ion cell, let's consider the problem of ion detection in an experiment involving (i) ion emission from a surface near the ion cell, i.e., laser desorption or $Cs^+$ ion SIMS, and (ii) ions entering the ion cell from an external ion source.

In the first case the ions are emitted from the surface in all directions and with a range of kinetic energies. Thus, the ions have initial velocities in the X-Y plane and along the Z-axis which are non-zero. In the absence of collisions with neutral atoms or molecules, the ions that are trapped in the ion cell will be randomly distributed with near thermal kinetic energies in all directions. This situation differs significantly from that for ions formed by electron impact ionization of neutral, thermal energy gas molecules. Ions formed by an electron beam will ideally result in the formation of a narrow ribbon of ions along the Z-axis corresponding to the path of the electron beam. The resulting distribution of ions will have negligible kinetic energies in the X-Y plane. The resulting ion motion can be effectively trapped by applying electrostatic trapping potentials along the Z-axis. Application of this trapping procedure to an ion population which has kinetic energies and therefore motions which are not limited to the Z-axis, no longer results in a coherent ion packet located in the center of the cell.

Ions formed in an external ion source and injected into the ICR cell are subject to effects which are analogous to the case of particle desorption ionization in the ICR cell. However, experimental results indicate limits in the resolution of high mass ions due to the finite duration of the time-domain image current signal. This observation is consistent

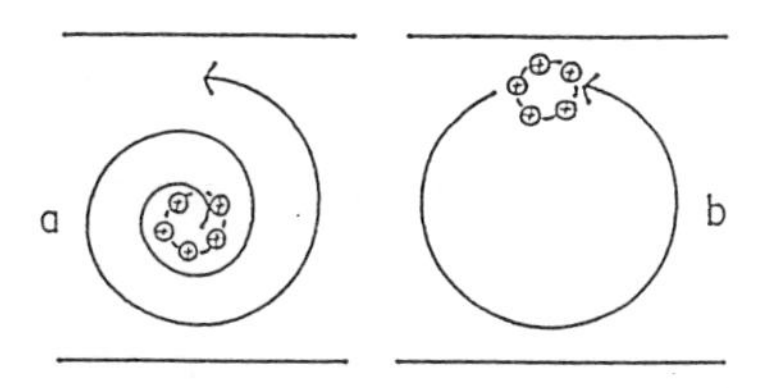


Figure 7. a) excitation of a coherent packet of ions b) a coherently cycloiding packet of ions which induces the image current signal.

Figure 8. Production of an image current signal from spatially random but phase coherent ion population.

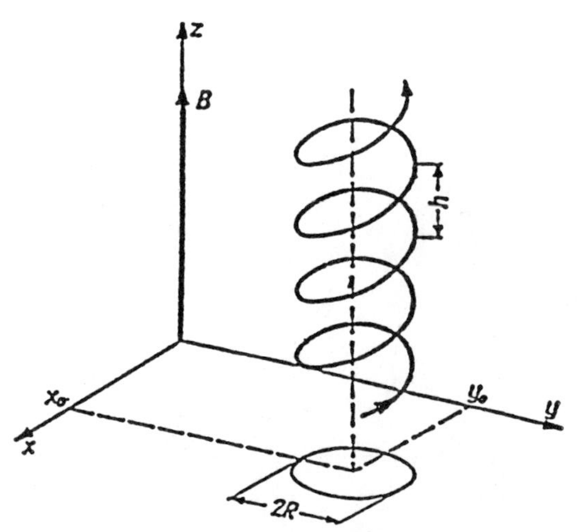


Figure 9. Ion path in a uniform magnetic field (a helical line of pitch h circumscribed on a cylinder of radius R).

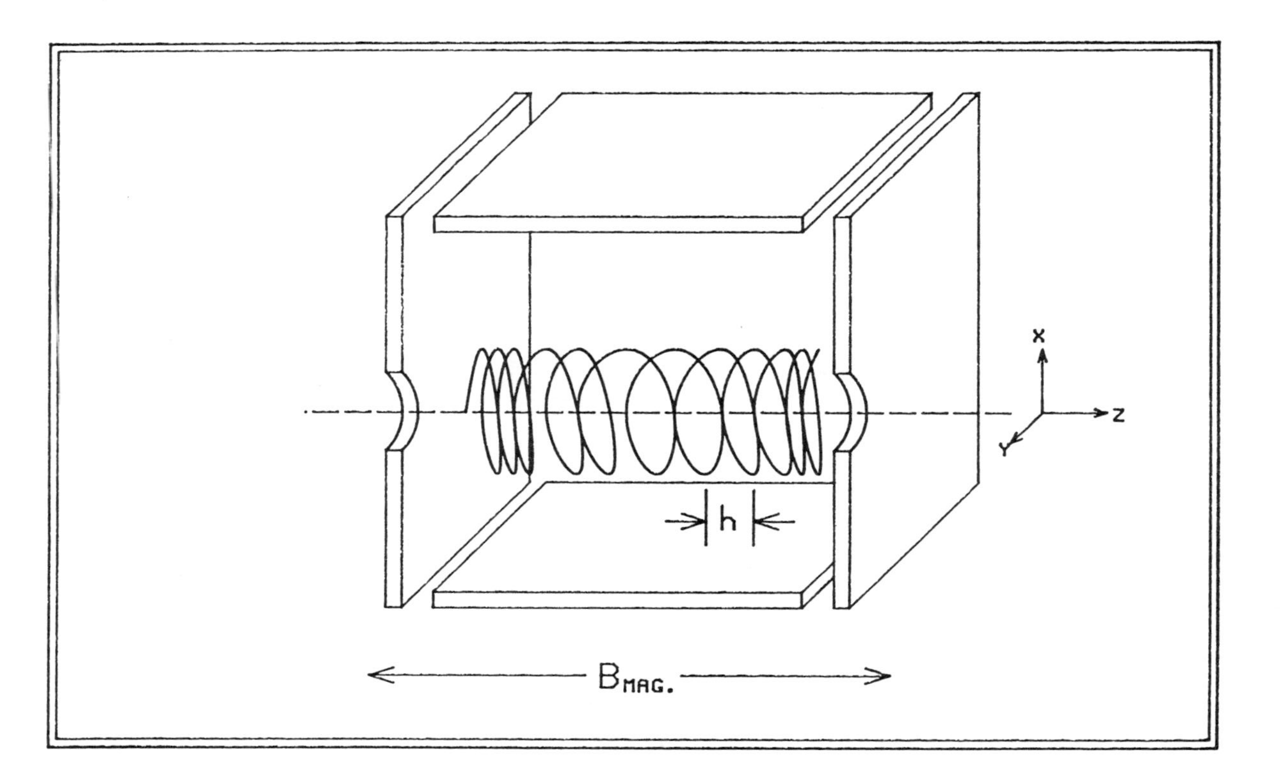


Figure 10. Variation of the pitch of the helical motion of ions along the Z-axis due to an imposed electrostatic field on the trapping plates.

with the inability of simple trapping procedures to produce a coherent packet of ions from the injected ion beam. In this case, where ions are produced external to the magnetic field and directed into the ion cell (either by a quadrupole or electrostatic ion guides) we must consider the effects of scattering of the ion beam in the X-Y plane by field inhomogeneities. Also, the secondary ion population produced will have a range of translational energies along the Z-axis. This distribution is caused by the post-ionization acceleration required to transport the secondary ions from the source to the ICR cell. The injected ions will acquire additional kinetic energies in the X-Y and/or Z directions by the combined effects of the inhomogeneities of the electrostatic and magnetic fields. Owing to the magnetic susceptibility of the stainless steel plates used to construct the ion cell, non-uniform magnetic field gradients in the X-Y plane will create magnetic inhomogeneities at the entrance orifices of the ion cell. Based on magnetic permeability of 304 and 316 stainless steel, induced magnetic field strengths on the order of 3,000 Gauss in a 3 Tesla magnet for unworked material and as much as 9,000 gauss for machined material are not unreasonable. Unlike the previous case where the ions are formed in the cell, ions produced externally must be transported through the magnetic perturbations prior to trapping. The trajectories of ions produced in the ICR cell with near zero velocities in the X-Y directions will not be strongly influenced by stray magnetic fields located at the extremes of the ICR cell, i.e., end trap plates. Owing to the magnitude of stray magnetic fields caused by the stainless steel material comprising the ICR cell, trajectories of ions being transported through the cell orifices will be deflected and appreciable gains in the kinetic energies in the X-Y plane will occur prior to trapping and detection. These effects can be modeled using ion trajectory calculations.[21] The calculated ion trajectories indicate that the magnetic field lines generated by the material comprising the ICR cell will redistribute the ions passing into the cell and that such effects are strongly dependant upon ion mass and kinetic energy along the Z-axis, and velocity components in the X-Y plane. Several different situations can be postulated to explain the loss of mass resolution at high mass for ions produced in an external ion source system.

The simplest case is that of an unfocused ion beam entering a weak magnetic field in the X-Y plane. This corresponds to a minimum magnitude of cross magnetic fields generated by unworked 304 stainless steel cell plates. Simulated ion trajectories for such a case are shown in Figure 11. Note the strong dependance on the mass of the ions entering the ion cell. Ions of m/z 50 are deflected in the X-Y plane by the cross magnetic fields, and the trajectory for m/z 5000 is virtually insensitive to the influence of the field and continues along the Z-axis. An interesting consequence of these results is that it suggests a potential mechanism for ion trapping from an external ion source, i.e., trapping by a magnetic bottleneck. A trapping mechanism of this type would be strongly mass dependant.

In the case of stronger magnetic fields, e.g., 304 stainless steel which has been machined, magnetic field gradients are imposed creating a stronger magnetic bottleneck at the orifices of the ion cell. Such a magnetic bottleneck will have strong effects on ions entering the cell close to the entrance orifice, accelerating them in the X-Y plane and reflecting them away from the orifice prior to entering the ICR cell. Only ions which are entering the magnetic bottleneck at the center of the orifice will

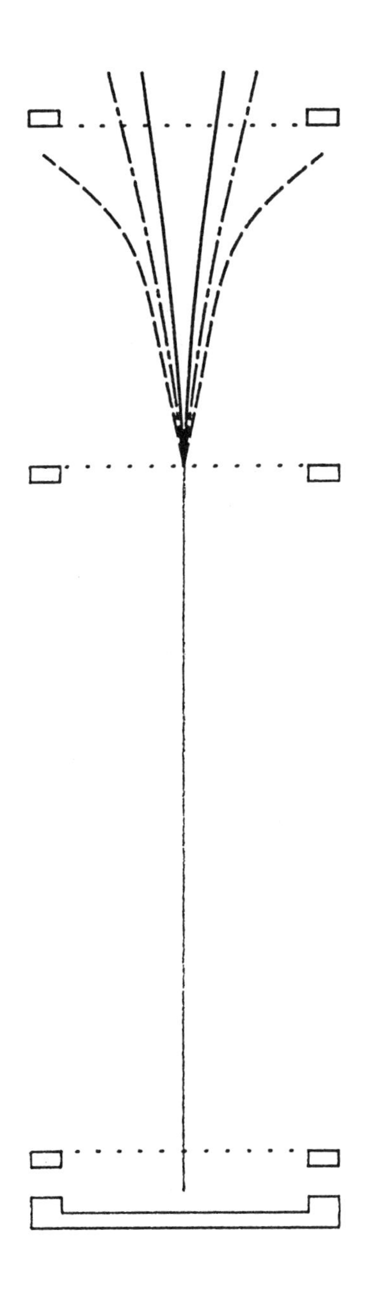

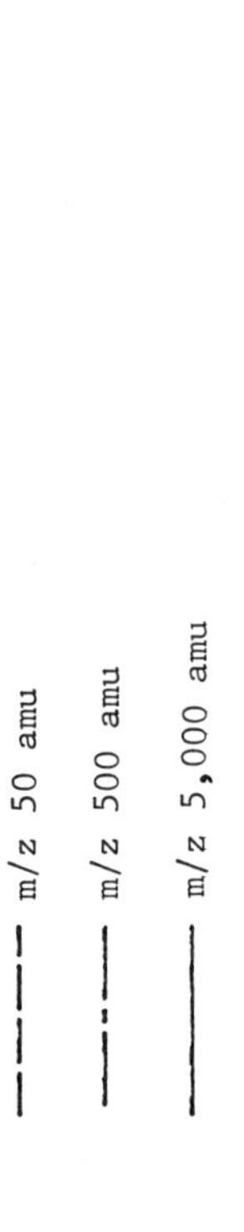


Figure 11. Effect of a 500 gauss cross magnetic field on the trajectories of an ion beam as a function of mass.

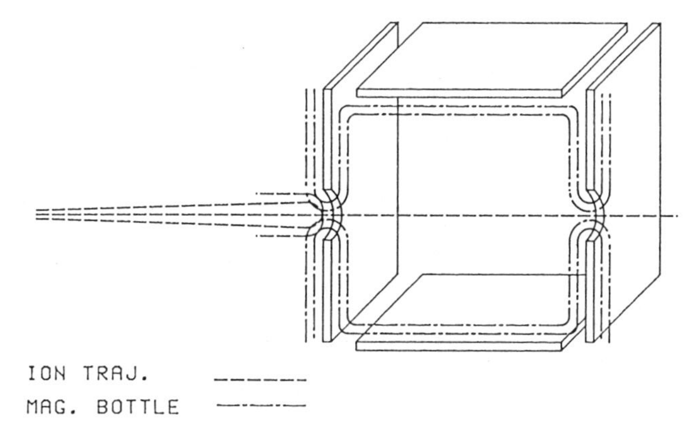


Figure 12. Effect of a magnetic bottleneck on diverging ion trajectories.

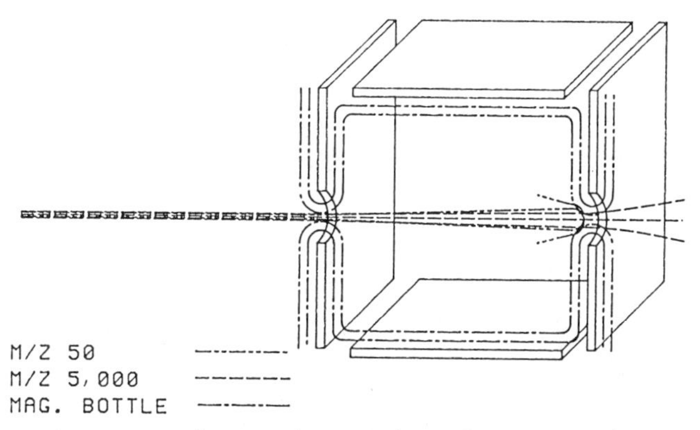


Figure 13. Magnetic trapping of ions in a magnetic bottle as a function of mass.

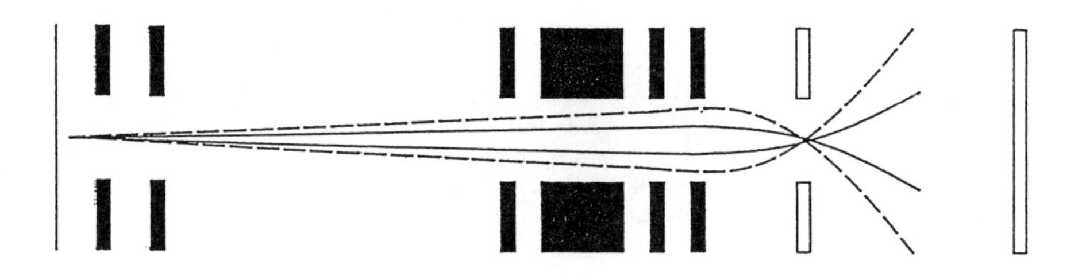


Figure 14. The effect of a cross magnetic field on the ion trajectories of a focused beam as a function of mass.

experience the weakest magnetic field effects and pass though the magnetic field into the ICR cell. As shown in Figure 12, ion trajectories which are diverging prior to entering the bottleneck will be deflected and not enter the cell. On the basis of these simulations we can predict that only ions having a narrow range of translational energies and angular trajectories will pass into the ion cell. This situation is consistent with experimental results reported by Smalley *et al.*[22], i.e., there is a narrow angle of acceptance for the trajectories of ions produced external to the magnetic field passing into the ICR cell. Owing to the existence of a second bottleneck surrounding the exit orifice on the rear trap plate, ions of low mass which are deflected in the X-Y plane by passing through the magnetic gradient of the entrance bottleneck will no longer have acceptable angular trajectories to pass through the exit orifice (Figure 13). This results in low mass ions being trapped magnetically in the ICR cell, whereas, high mass ions are less sensitive to the cross magnetic field gradients and therefore exit the ion cell prior to electrostatic trapping.

A possible solution to this problem is the use of electrostatic focusing lenses which bring the ion beam to a point focus at the entrance of the ICR cell. This electrostatic focusing technique converges all the ions at the entrance of the magnetic bottleneck maximizing the transmission of ions into the cell. After entering the cell, the focused beam diverges supplying sufficient X-Y velocity components to prevent the ions from exiting the rear magnetic bottleneck. Figure 14 shows the ion trajectory for a focused ion beam entering the ICR cell, diverging and being reflected back into the cell prior to electrostatic trapping. Based on these considerations, effective trapping and subsequent detection of ions with these types of trajectories cannot be achieved by simple electrostatic methods. Trapping of such ions must then be a function of both electrostatic and magnetic factors.

Experiments are presently underway to test the ideas presented above. Although the experiments are not yet completed, the predicted effects of magnetic inhomogeneities on ion motion and ion trapping are consistent with the available experimental data.[23] In addition, detailed measurements of ion partitioning in a two-section cell are consistent with the idea of magnetic inhomogeneities having very large effects on ion trapping and ion detection. In these experiments, ion partitioning in an ICR cell made of stainless steel plates resulted in no detectable signal from the transported ions. In order to investigate the effect of magnetic inhomogeneities on the transported ions, the stainless steel trap plates were replaced with trap plates made out of oxygen-free copper due to it's near zero magnetic susceptibility. Once the magnetic inhomogeneities were removed by replacing the stainless steel with oxygen-free copper, 50% of the ions produced in the source region were trapped and detected in the analyzer region after allowing equal partitioning between the two regions. Based on these results, the presence of magnetic inhomogeneities caused by the magnetic susceptibility of stainless steel deflects the ions being transported between the two regions of the dual cell, leading to quenching of the ions prior to detection.

## Literature Cited

(1) (a) M.B. Comisarow and A.G. Marshall Chem. Phys. Lett. **1974**, 25, 282;

(b) M.B. Comisarow and A.G. Marshall J. Chem. Phys. **1976**, 64, 110.
(2) M.L. Gross and D.L. Rempel Science **1984**, 226, 261.
(3) C.L. Wilkins and M.L. Gross Anal. Chem. **1981**, 53, 1661A.
(4) D.A. McCrery, E.B. Ledford, and M.L. Gross Anal. Chem. **1982**, 54, 1437.
(5) M.E. Castro and D.H. Russell Anal. Chem. **1984**, 56, 578.
(6) M. Allemann, Hp. Kellerhals, and K.P. Wanczek Int. J. Mass Spectrom. Ion Phys. **1983**, 46, 139.
(7) R.T. McIver, R.L. Hunter, and W.D. Bowers Int. J. Mass Spectrom. Ion Proc. **1985**, 64, 67.
(8) D.F. Hunt, J. Shabanowitz, R.T. McIver, R.L. Hunter, J.E.P. Syka Anal Chem **1985**, 57, 765.
(9) S. Ghaderi and D. Littlejohn "Proceedings of the 33rd Annual Conference on Mass Spectrometry and Related Topics", San Diego, California, May 26-31, **1985**, pp. 727-728
(10) P. Kofel, M. Allemann, Hp. Kellerhals, K.-P. Wanczek Int. J. Mass Spectrom. Ion Proc. **1985**, 65, 97.
(11) D.F. Hunt, J. Shabanowitz, J.R. Yates, N.-Z. Zhu, M.E. Castro, and D.H. Russell Proc. Natl. Acad. Sci. **1987**, 84, 620.
(12) D.H. Russell and M.E. Castro, in "Mass Spectrometry in Biomedical Research" S.J. Gaskell (Ed.), J. Wiley and Sons, New York, **1986**, pp 313-338;
(b) D.H. Russell Mass Spectrom. Rev. **1986**, 5, 167.
(13) B.T. Chait and F.H. Field J. Am. Chem. Soc. **1984**, 106, 1931.
(14) B.T. Chait and F.H. Field Int. J. Mass Spectrom. Ion Proc. **1985**, 65, 169.
(15) I.J. Amster, F.W. McLafferty, M.E. Castro, D.H. Russell, R.B. Cody, and S. Ghaderi Anal. Chem. **1986**, 58, 483.
(16) W. Aberth Anal. Chem. **1986**, 58, 1221.
(17) D. Schuch, K.M. Chung, H. Hartmann Int. J. Mass Spectrom. Ion Proc. **1984**, 56, 109.
(18) P. Kofel, M. Allemann, Hp. Kellerhals, K.-P. Wanczek Int. J. Mass Spectrom. Ion Proc., **1986**, 74, 1.
(19) F.H. Laukien Int. J. Mass Spectrom. Ion Proc. **1986**, 73, 81.
(20) B. Paszkowski, "Electron Optics", English Translation by R.C.G. Leckey, Iliffe Books, LTD.,New York, **1968**, Chapters 1 and 7.
(21) SIMION, "Simulated Ion Trajectories" is a computer program developed by D. McGilvery, LaTrobe University. This Program has been refined and expanded by the group (D.A. Dahl and J.E. Delmore) at Idaho National Laboratory, Department of Energy.
(22) J.M. Alford, P.E. Williams, D.J. Trevor, and R.E. Smalley Int. J. Mass Spectrom. Ion Proc. **1986**, 72, 33.
(23) E.L. Kerley, C.D. Hanson, M.E. Castro, and D.H. Russell, to be submitted to Anal. Chem.

RECEIVED June 15, 1987

Chapter 7

# Tandem Fourier Transform Mass Spectrometry of Large Molecules

**Fred W. McLafferty, I. Jonathan Amster, Jorge J. P. Furlong, Joseph A. Loo, Bing H. Wang, and Evan R. Williams**

**Chemistry Department, Cornell University, Ithaca, NY 14853-1301**

Structural characterization of increasingly large molecules requires exponentially increasing amounts of molecular information as well as separation methods of higher resolution. Tandem mass spectrometry using the Fourier-transform instrument is promising for both of these requirements as described in other papers of this book. Helpful advances summarized here are the use of secondary ion mass spectrometry and $^{252}Cf$ plasma desorption for ionization of non-volatile molecules, and the formation of MS-II spectra using the surface induced dissociation method of Cooks, the electron dissociation method of Cody and Freiser, and neutralization of multiply protonated ions to form unstable hypervalent species. The use of the Hadamard transform technique reduces the time required for recording n MS-II spectra by 4/n.

In the last decade mass spectrometry (MS) has been revolutionized by new methods for ionizing large molecules, even as large as trypsin, molecular weight (MW) 23,463 (1). Increasing the size of an unknown molecule results in an exponential increase in the structural information needed for its identification (2-4). MS is already well recognized for the unusual amount of information obtainable from picomole samples, and high resolution and tandem MS (MS/MS) (5) can provide substantially more information. However, conventional magnetic sector instruments must scan the mass spectrum, expending sample continuously although measuring only a very small fraction of the mass scale at any one time. Large molecules exacerbate this problem, increasing the length of the mass scale. Further, for high resolution measurements the fraction of the mass scale measured at any time must be much smaller. For MS/MS these scanning problems are increased exponentially, as measurement of MS-II spectra can be necessary for a multiplicity of primary ion species separated by MS-I (6). Multichannel detection is an obvious solution to this problem. Array

0097-6156/87/0359-0116$06.00/0

detectors can only be used with plane-of-focus magnetic sector instruments (7), seriously limiting the mass range covered in most instruments. The time-of-flight (TOF) instrument is used in many large molecule studies because it achieves multichannel recording over a virtually unlimited mass range. Offsetting this (1, 8-13), TOF resolution of >1000 has been difficult to achieve (although >10,000 has been reported recently) (14, 15), and to date TOF has exhibited poor capabilities for MS/MS (16). Thus Fourier-Transform (FT) MS (17-23), as detailed in the accompanying articles, has unique promise for characterization of minimum amounts of large molecules, with its multiple capabilities for MS/MS and multichannel detection of ions over a wide mass range. Further, unusually high resolution is possible at low masses. This paper focuses on special ionization methods such as secondary ion MS (SIMS) (12, 13, 24-28) and $^{252}Cf$ plasma desorption (PD), and on MS/MS methods for characterizing primary ions, such as surface induced dissociation (SID), laser photodissociation, and neutralization of multiply charged ions. A Hadamard transform method for more efficient recording of multiple MS-II spectra is also proposed.

## FTMS Ionization Methods

Multichannel recording or ion storage capabilities, both available with FTMS, are virtually required for some ionization methods, such as those producing intermittent ion pulses (e.g., laser desorption, LD) (29-31) or low ion currents (e.g., PD). On the other hand, FTMS has much lower pressure requirements ($<10^{-8}$ torr), especially for high resolution measurements, so that exterior differentially-pumped (23, 32, 33) or dual cell (20, 34) ion sources are beneficial for most ionization methods with FTMS. Trapping the ions in the FTMS cell for measurement is now recognized as a key problem, although the parameters are poorly understood. For this, generating ions at the cell entrance, such as by LD, would appear to be advantageous; however, Hunt and co-workers (23) have reported molecular ion species of cytochrome-C (MW 12,364), the highest mass organic ions recorded to date by FT, using fast atom bombardment (FAB) ionization in an exterior source. Two other methods, SIMS and PD, have been shown to have highly promising attributes, mainly using the TOF instrument; their applicability with FTMS has been recently demonstrated. Our recent results indicate that using a 193 nm excimer laser for LD offers a useful alternative to the infrared laser used previously (29-31).

SIMS. Russell and co-workers have used SIMS/FTMS with 4 keV $Cs^+$ ions to form dimer ions of vitamin B-12 (25) and molecular ions of β-cyclodextrin (*m/z* 1135) (26). With them and Nicolet scientists we have measured $Cs_{63}I_{62}^+$, *m/z* 16,241, with a FTMS instrument with a 3 tesla magnet (35). Using a 7 tesla magnet Hunt and coworkers recently achieved more than double this mass value for $(CsI)_nCs^+$ ions using the related FAB technique (23). Amster and co-workers (28) in an extensive study have shown the general applicability of SIMS/FTMS using higher energy (11 keV) $Cs^+$ ions. In comparison to the much more widely used method FAB, subpicomole sensitivities are possible, and information on both molecular weight and structure from fragment ions is available (Figures 1 and 2) without the ubiquitous background peaks from the FAB liquid matrix. However, under present operating conditions sensitivity falls off rapidly for masses above ~2000 daltons, a problem which is addressed by Gross, Hunt, Russell, and coworkers in separate articles in this volume. Convenient sources capable of forming 12 keV $U^{5+}$ have been described (36); our preliminary results using multiply charged primary ions for SIMS are encouraging.

$^{252}$Cf-PD. An ionization method which is unique in producing essentially no emitted gases, and thus no pressure problem for FTMS, is $^{252}$Cf-PD (1, 8-11, 37-39). Its use with a nitrocellulose sample substrate also gives abundant multiply charged ions, even $(M + 6H)^{6+}$ from trypsin, greatly extending the mass (*m*) range of instruments like FTMS with an upper *m/z* (*z* = number of charges) limit. Recently, Tabet, Gaumann, and coworkers showed the potential of $^{252}$Cf PD ionization with FTMS, reporting $(M + K)^+$, *m/z* 574, from leu-enkephalin (37). Some more recent FTMS studies (22, 38) do not report molecular ions from compounds this large. Loo, Chait, and coworkers find PD/FTMS to be generally applicable up to *m/z* 2000, producing abundant $(M + A)^+$ ions (A = H, Li, Na, K, Ca) and structurally-informative fragment ions (39). Key experimental techniques developed in the pioneering work of Macfarlane (8), Sundqvist (1), Cotter (11), and others are used; samples are evaporated from solution onto nitrocellulose that has been electrosprayed onto a mylar film, or the samples are electrosprayed directly onto the mylar from a solution that is 1:1 (molar) with glutathione. The mylar is placed on top the 50 µcurie $^{252}$Cf source mounted on the end of the FTMS sample probe, so that when introduced into the vacuum system the sample is ~3 mm from the ion cell entrance on the magnetic field axis. Data collection is started when the pressure reaches ~$10^{-8}$ torr; instrument steady-state pressure is ~$2x10^{-9}$ torr, essentially the same with or without the $^{252}$Cf and sample. The potential on the cell trapping plates and aluminized mylar sample film is critical; the total ion signal increases rapidly while increasing the potential to 5 V, and more slowly to 10 V, although the abundance of the molecular ion species maximizes around 5 V. Optimum sensitivity was obtained by allowing ions to build up in the cell for 1 or 2 minutes (more for larger molecules) followed by RF excitation and 65 ms transient signal measurement; subsequent ion removal from the cell ("quenching") decreased sensitivity by at least an order of magnitude. Co-adding signals for at least 14 hours improved signal/noise of the spectrum and the relative abundance of the molecular ion species, indicating that lifetime is not a problem with these large ions. Unfortunately, under these conditions a sample of insulin gave no significant signal, reflecting our current problem with $Cs^+$ SIMS for higher molecular weight samples.

## Methods for Producing MS-II Spectra

For MS/MS characterization the MS-I separated ions must be made to undergo dissociation or other reactions. The most widely used

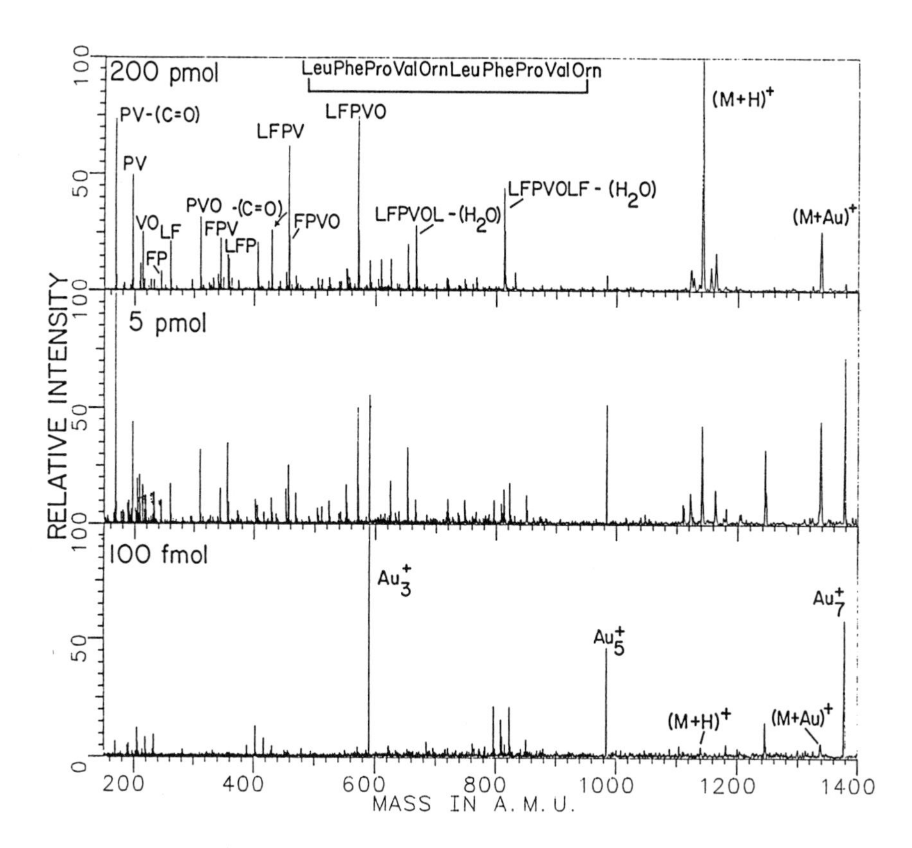


Figure 1. SIMS/FTMS spectrum of gramicidin S and glutathione (1:2) on a gold target. Total samples deposited were as follows: $2 \times 10^{-10}$ mol (top), $5 \times 10^{-12}$ mol (middle), $10^{-13}$ mol (bottom). (Reproduced from ref. 28. Copyright 1987 American Chemical Society.)

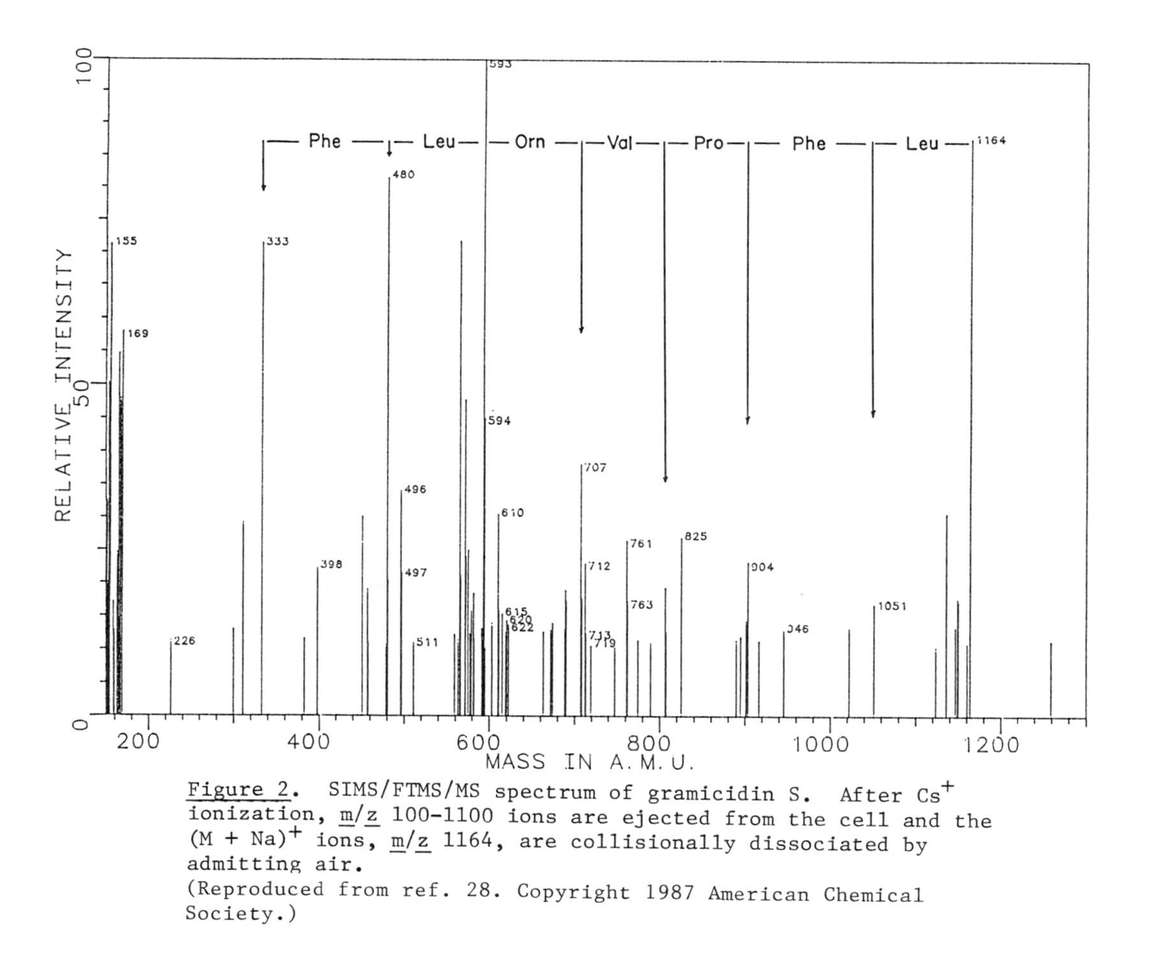


Figure 2. SIMS/FTMS/MS spectrum of gramicidin S. After $Cs^+$ ionization, m/z 100-1100 ions are ejected from the cell and the $(M + Na)^+$ ions, m/z 1164, are collisionally dissociated by admitting air.
(Reproduced from ref. 28. Copyright 1987 American Chemical Society.)

method to date is collisionally activated dissociation (CAD) (40-43). For energy-dependent mass analyzers CAD spectra have the disadvantage that translational energy from dissociation spreads the kinetic energy of the resulting ions, lowering resolution (vide supra). FTMS mass analysis is energy-independent, avoiding this problem. Multikilovolt grazing collisions have the advantage that relative abundances of ions from higher energy processes are nearly independent of precursor ion internal energy (41, 43, 44), but these energies cannot be achieved in FTMS for larger (a few hundred daltons) ions. CAD of 10-100 eV kinetic energy ions with efficient product collection over large angles (possible with triple quadrupole and FTMS instrumentation) has the advantage that the substantial effect of collision energy on product ion abundances provides an extra parameter for characterization (42, 45).

Surface Induced Dissociation. CAD has the basic FTMS disadvantage of requiring a collision gas and thus increasing the ion source pressure. This problem can be reduced by pulsed valve introduction of the collision gas (46), and eliminated by using the exciting technique of Surface Induced Dissociation (SID), recently introduced by Cooks (47, 48). SID also has the unusual advantage of being a convenient means of adding a variable, and relatively narrow, range of internal energies to the ion. Preliminary work by our laboratory (49) has effected SID by accelerating ions collected in one side of the dual cell through the conductance limit orifice to strike a metal surface in the other cell. Results for toluene molecular ions using different accelerating energies are similar to those reported by Cooks (47), and dissociation efficiencies appear to be relatively high.

Other Activation. Efficient (>10%) CAD of positive ions appears possible using collisions with electrons (50). Recent results by Cody and Freiser are very promising (51), confirmed by initial experiments with our FTMS instrumentation (49). Laser photodissociation appears to be particularly applicable for FTMS ion characterization, with 193 nm pulses from an excimer laser giving fragmentation (with conversion yields as high as 25%) that appears to be associated with specific absorption and dissociation at groups such as phenylalanine having high absorptivities (52). Recent results by Hunt and coworkers described in this volume, confirmed in our laboratory, indicate that this is a very promising method for ions up to mass 2000 using 193 nm laser radiation.

Activation of Large Ions. The multikilovolt CAD spectra of larger ions ($m/z$ >~1500) can be dominated by the loss of small neutral species such as $H_2O$ and $NH_3$ (53, 54), consistent with theoretical predictions that bond cleavages at molecular extremities will be increasingly favored over central cleavages in such "fender-bender" collisions of increasingly large ions (55). Increasing amounts of added internal energy are necessary to effect ergodic dissociation because of the increasing number of degrees of freedom among which the added energy must be distributed. Recent studies by Biemann with an elegant tandem double-focusing mass spectrometer (56) confirm that CAD of large peptide ions becomes very inefficient above about mass 2500. As reported in this volume, Russell has shown that higher mass collision gases are more effective because they carry away less of the collision energy.

We have proposed (57) an additional method for such fast dissociation using ions such as (peptide + $H_3)^{3+}$ (eq 1). Our studies have shown that neutralization of an ionized saturated heteroatom site such as a quaternary ammonium ion gives an unstable hypervalent species.

$$RNHR_1NHR_2NHR_3 + 3H^+ \rightarrow R\text{-}N^+H_2\text{-}R_1\text{-}N^+H_2\text{-}R_2\text{-}N^+H_2\text{-}R_3 \quad (1)$$

$$+ e^- \rightarrow R\text{-}N^+H_2\text{-}R_1\text{-}N{*}H_2\text{-}R_2\text{-}N^+H_2\text{-}R_3 \quad (2)$$

$$\rightarrow R\text{-}N^+H_2\text{-}R_1\cdot + H_2N\text{-}R_2\text{-}N^+H_2\text{-}R_3 \quad (3)$$

Thus the multiply protonated species, $R^+CO\text{-}N^+H_2\text{-}R'^+$, on neutralization (eq 2) of the central charge site could yield (eq 3) the singly charged odd- and even-electron products $R^+\text{-}CO^\cdot + H_2N\text{-}R'^+$. The effectiveness could be reduced by competing proton loss or stabilization by zwitter ion formation, $\rightarrow R^+\text{-}CO^-\text{-}N^+H_2\text{-}R'^+$. However, neutralization with electrons (effected as proposed above for electron CAD) would deposit vibrational energy corresponding to the ionization energy (~5 eV) of the new neutral site, enhancing its non-ergodic dissociation.

The selected primary ions can also be characterized by ion molecule reactions with the FTMS-2000 dual cell and/or surface induced reactions (48). Reactions specific for particular functionalities can be used to count the number of these in a primary ion such as the number of hydrogen atoms that can be exchanged with a deuterium-containing species (59). Further, their position may well be distinguished by the reaction's effect on fragment ions in MS-II spectra.

## Hadamard Transform Enhancement in Measurement

For tandem mass spectrometry the multichannel capability of dissociating any combination of parent ions to form a collective MS-II spectrum of daughter ions, possible on Fourier-transform and ion trap instruments, can be used to increase the signal/noise and/or speed of data collection (60). A Hadamard transform advantage of $0.5\, n^{0.5}$ in signal/noise or 4/n in required time is possible in measuring individual MS-II spectra of n parent ions; n collective spectra using different combinations of 0.5n of the parent ions are measured, and the individual contributions at each specific mass in each MS-spectrum is calculated from the n simultaneous equations representing the summed intensity values at each mass. For MS/MS detection of many target compounds which occur infrequently, "no peak" information can rapidly reduce the number of possible target compounds present (49). Any peak in the MS-II spectrum of a target parent ion which is negligible in the collective MS-II spectra of all parent ions shows that the corresponding target compound is negligible (60).

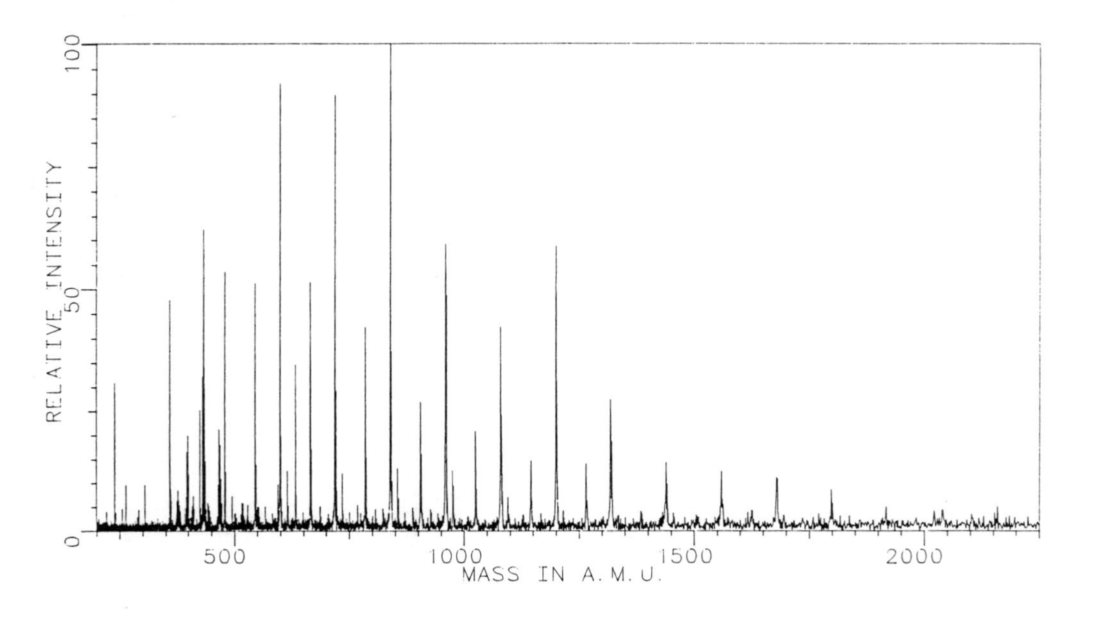


Figure 3. PD/FTMS negative ion spectrum of poly(butene-1-sulfone) from bombardment by 100 MeV fission fragments from $^{252}Cf$. Major ion series separated by 120 daltons ($C_4H_8$, 56; $SO_2$, 64) correspond to $CH_3CH{=}CHCH_2(SO_2C_4H_8)_nSO_2^-$ and $HO_2SCH(C_2H_5)CH_2(SO_2C_4H_8)_nSO_2H$ (Reproduced from ref. 61. Copyright 1987 American Chemical Society.)

Degradation Mechanisms of E-Beam Resists. Research supported by IBM has the objective of elucidating the reaction mechanisms from high energy radiation of poly(olefin-sulfones). These have by far the highest sensitivity (G value) as E-beam resist polymers. The formation of submicron-dimension images in such resists with focused high energy electrons is used to produce microcircuits. From analysis of the condensed products of the radiation degradation an "unzipping" (depolymerization) mechanism was proposed. Our preliminary study of the primary products from bombardment of poly(butenesulfone) by 10 keV $Cs^+$ ions, or by $^{252}Cf$ fission products (Figure 3), in the high vacuum FTMS cell indicates an entirely different mechanism. The spectrum products can be explained by a radical-chain reaction propagated across the polymer chains utilizing pre-formed hydrogen bonds, consistent with the high radiation sensitivity (61).

## Acknowledgments

The authors are indebted to B. T. Chait, R. B. Cody, Jr., R. J. Cotter, F. H. Field, T. Gäumann, S. Ghaderi, D. F. Hunt, D. P. Littlejohn, R. D. Macfarlane, C. J. McNeal, P. Roepstorff, B. U. R. Sundqvist, and J. C. Tabet for helpful advice, and to the National Institutes of Health (Grant GM16609) and the Army Research Office (Grant DAAL03-86-K-0088) for generous financial support.

## Literature Cited

1. Jonsson, G. P.; Hedin, A. B.; Håkansson, P. L.; Sundqvist, B. U. R.; Säve, B. G. S.; Nielsen, P. F.; Roepstorff, P.; Johansson, K.-E.; Kamensky, I.; Lindberg, M. S. L. Anal. Chem. 1986, 58, 1084-1087.
2. McLafferty, F. W. Interpretation of Mass Spectra, Third Edition; University Science Books: Mill Valley, CA, 1980.
3. Facchetti, S., Ed. Mass Spectrometry of Large Molecules; Elsevier: Amsterdam, 1985.
4. Burlingame, A. L.; Castagnoli, N., Jr., Eds., Mass Spectrometry in the Health and Life Sciences; Elsevier: Amsterdam, 1985.
5. McLafferty, F. W., Ed., Tandem Mass Spectrometry; John Wiley: New York, 1983.
6. Coutant, J. E.; McLafferty, F. W. Int. J. Mass Spectrom. Ion Phys. 1972, 8, 323-339.
7. Boerboom, A. J. H. In ref. 5, Chapter 11.
8. Macfarlane, R. D. Anal. Chem. 1983, 55, 1247A-1264A.
9. Chait, B. T.; Field, F. H. J. Am. Chem. Soc. 1984, 106, 1931-1938.
10. Pannell, L. K.; Sokoloski, E. A.; Fales, H. M.; Tate, R. L. Anal. Chem. 1985, 57, 1060-1067.
11. Alai, M.; Demirev, P.; Fenselau, C.; Cotter, R. J. Anal. Chem. 1986, 58. 1303-1307.
12. Benninghoven, A., Ed. Ion Formation from Organic Solids; Springer-Verlag: Berlin, 1983.

13. Benninghoven, A.; Niehuis, E.; Friese, T.; Greifendorf, D.; Steffens, P. Org. Mass Spectrom. 1984, 19, 346.
14. Benninghoven, A.; Wollnik, H., Private Communications, 1986.
15. Walter, K.; Boesl, U.; Schlag, E. W. Int. J. Mass Spectrom. Ion Processes, submitted.
16. Haddon, W. F.; McLafferty, F. W. Anal. Chem. 1969, 41, 31-36.
17. Comisarow, M. B.; Marshall, A. G. Chem. Phys. Lett. 1974, 25, 282-283.
18. Gross, M. L.; Rempel, D. L. Science 1984, 226, 261-268.
19. Marshall, A. G. Acc. Chem. Res. 1985, 18, 316-322.
20. Cody, R. B., Jr.; Kinsinger, J. A.; Ghaderi, S.; Amster, I. J.; McLafferty, F. W.; Brown, C. E. Anal. Chim. Acta 1985, 178, 43-66.
21. Laude, D. A., Jr.; Johlman, C. L.; Brown, R. S.; Weil, D. A.; Wilkins, C. L. Mass Spectrom. Rev. 1986, 5, 107-166.
22. Russell, D. H. Mass Spectrom. Rev. 1986, 5, 167-189.
23. Hunt, D. F.; Shabanowitz, J.; Yates, J. R., III; Zhu, N-Z; Russell, D. H.; Castro, M. E. Proc. Natl. Acad. Sci. U.S.A., 1987, 84, 620-623.
24. Lange, W.; Jirikowsky, M.; Benninghoven, A. Surf. Sci. 1984, 136, 419-436.
25. Castro, M. E.; Russell, D. H. Anal. Chem. 1984, 56, 578-581.
26. Castro, M. E.; Mallis, L. M.; Russell, D. H. J. Am. Chem. Soc. 1985, 107, 5652-5657.
27. Wandass, J. H.; Gardella, J. A. J. Am. Chem. Soc. 1985, 107, 6192-6195.
28. Amster, I. J.; Loo, J. A.; Furlong, J. J. P.; McLafferty, F. W. Anal. Chem. 1987, 59, 313-317.
29. McCrery, D. A.; Ledford, E. G.; Gross, M. L. Anal. Chem. 1982, 54, 1435-1437.
30. Cotter, R. J.; Tabet, J.-C. Int. J. Mass Spectrom. Ion Phys. 1983, 53, 151.
31. Cody, R. B.; Amster, I. J.; McLafferty, F. W. Proc. Natl. Acad. Sci. U.S.A. 1985, 82, 6367-6370.
32. Hunt, D. F.; Shabanowitz, J.; Yates, J. R., III; McIver, R. T., Jr.; Hunter, R. L.; Syka, J. E. P.; Amy, J. Anal. Chem. 1985, 57, 2728-2733.
33. Kofel, P.; Allemann, M.; Kellerhals, H. P.; Wanczek, K. P. Adv. Mass Spectrom. 1986, 10, 885-886.
34. Ghaderi, S.; Littlejohn, D. P. Adv. Mass Spectrom. 1986, 10, 875.
35. Amster, I. J.; McLafferty, F. W.; Castro, M. E.; Russell, D. H.; Cody, R. B., Jr.; Ghaderi, S. Anal. Chem. 1986, 58, 483-485.
36. Brown, I. G.; Galvin, J. E.; Gavin, B. F.; MacGill, R. A. Rev. Sci. Instrum. 1986, 57, 1069-1084.
37. Tabet, J. C.; Rapin, J.; Poretti, M.; Gaumann, T. Chimia 1986, 40, 169-171.
38. Viswanadham, S. K.; Hercules, D. M.; Weller, R. R.; [illegible]am, C. S. Biomed. Environ. Mass Spectrom. 1987, 14, 43-45.
39. Loo, J. A.; Williams, E. R.; Amster, I. J.; Furlong, J. J. P.; Wang, B. H.; McLafferty, F. W.; Chait, B. T.; Field, F. H. Anal. Chem., accepted.
40. McLafferty, F. W.; Bente, P. F., III; Kornfeld, R.; Tsai, S.-C.; Howe, I. J. Am. Chem. Soc. 1973, 95, 2120.

41. McLafferty, F. W.; Kornfeld, R.; Haddon, W. F.; Levsen, K.; Sakai, I.; Bente, P. F., III; Tsai, S.-C.; Schuddemage, H. D. R. J. Am. Chem. Soc. 1973, 95, 3886-3892.
42. Dawson, P. H.; Douglas, D. J. In ref. 5, Chapter 6.
43. Todd, P. J.; McLafferty, F. W. In ref. 5, Chapter 7.
44. McLafferty, F. W.; Proctor, C. J. Org. Mass Spectrom. 1983, 18, 272.
45. McLuckey, S. A.; Cooks, R. G. In ref. 5, Chapter 15.
46. Carlin, T. J.; Freiser, B. S. Anal. Chem. 1983, 55, 571-574.
47. Mabud, M. A.; DeKrey, M. J.; Cooks, R. G. Int. J. Mass Spectrom. Ion Proc. 1985, 67, 285-294.
48. DeKrey, M. J.; Kenttamaa, H. I.; Wysocki, B. H.; Cooks, R. G. Org. Mass Spectrom. 1986, 21, 193-195.
49. McLafferty, F. W.; Williams, E. R. Proc. 1986 Conf. Chemical Defense Res., Aberdeen Proving Ground, MD, 1987.
50. Cody, R. B.; Freiser, B. S. Anal. Chem. 1979, 51, 547-551.
51. Cody, R. B.; Freiser, B. S. Anal. Chem. 1987, 59, 1054-1056.
52. Bowers, W. D.; Delbert, S.-S.; McIver, R. T., Jr. Anal. Chem. 1986, 58, 969-972.
53. Amster, I. J.; Baldwin, M. A.; Chang, M. T.; Proctor, C. J.; McLafferty, F. W. J. Am. Chem. Soc. 1983, 105, 1654-1655.
54. Neumann, G. M.; Derrick, P. J. Org. Mass Spectrom. 1984, 19, 165-170.
55. Bunker, D. L.; Wang, F.-M. J. Am. Chem. Soc. 1977, 99, 7457-7459.
56. Biemann, K. Anal. Chem. 1986, 58, 1288A-1300A.
57. McLafferty. F. W. In Mass Spectrometry in the Analysis of Large Molecules; McNeal, C. J., Ed.; John Wiley: New York, 1986, pp 107-120.
58. Wesdemiotis, C.; Danis, P. O.; Feng, R.; Tso, J.; McLafferty, F. W. J. Am. Chem. Soc. 1985, 107, 8059-8066.
59. Ingemann, S.; Nibbering, N. M. M. J. Org. Chem. 1983, 48, 183-191.
60. McLafferty, F. W.; Stauffer, D. B.; Loh, S. Y.; Williams, E. R., Anal. Chem., accepted.
61. Loo, J. A.; Wang, B. H.; Wang, F. C.-Y.; McLafferty, F. W.; - Klymko, P. Macromolecules 1987, 20, 698-700.

RECEIVED May 12, 1987

# Chapter 8

# Analytical Applications of Laser Desorption–Fourier Transform Mass Spectrometry for Nonvolatile Molecules

**R. S. Brown and C. L. Wilkins**

**Department of Chemistry, University of California, Riverside, CA 92521**

Laser desorption Fourier transform mass spectrometry (LD-FTMS) results from a series of peptides and polymers are presented. Successful production of molecular ions of peptides with masses up to ~2000 amu is demonstrated. The amount of structurally useful fragmentation diminishes rapidly with increasing mass. Preliminary results of laser photodissociation experiments in an attempt to increase the available structural information are also presented. The synthetic biopolymer poly(phenylalanine) is used as a model for higher molecular weight peptides and produces ions approaching m/z 4000. Current instrument resolution limits are demonstrated utilizing a poly(ethyleneglycol) polymer, with unit mass resolution obtainable to almost 4000 amu.

Laser desorption Fourier transform mass spectrometry (LD-FTMS) has emerged an an alternative soft ionization method for samples which are difficult to analyze by more conventional mass spectrometric methods. Since its initial demonstration (1) for organic samples, there has been steady progress in applications of LD-FTMS to a wide variety of samples. These have included such diverse substances as simple peptides (2,3), sugars (4), polymers (2,5,6), porphyrins (7,8) and oligosaccharides (9,10) of various molecular weights. Although still a comparatively new technique, for many samples LD-FTMS compares favorably with more established ionization methods such as fast atom bombardment (FAB) and field desorption (FD), the former having become the method of choice for most non-volatile polar samples. LD-FTMS offers several advantages over FAB, including the lack of need for a liquid matrix, which is a source of high chemical background in the FAB method. Additionally, the absence of a liquid matrix removes the necessity that the sample be soluble. Indeed, insoluble samples can be applied directly in solid form to a sample probe and analyzed by LD-FTMS with good results (6). Many organometallics, metalloporphyrins and polymers do not yield spectra when examined by

0097-6156/87/0359-0127$06.00/0

FAB-MS, due to their insolubility in the liquid matrix. Thus, LD-FTMS should be considered complementary to FAB-MS.

Due to the relatively small number of laboratories equipped with LD-FTMS, the method is not as well known or as highly developed as FAB ionization which is now available for most commercial magnetic sector and quadrupole instruments. The pulsed nature of the lasers typically employed are ideally matched with mass spectrometers capable of simultaneous detection or rapid analysis. Thus FTMS or time of flight (TOF) instruments are best suited for LD applications. However, the latter possesses limited mass resolution. This chapter will attempt to present the current capabilities and limitations of LD-FTMS and to address its future potential.

## Experimental

The general experimental arrangement for LD-FTMS has been presented in detail elsewhere (2) and only a brief description will be provided here. For sample desorption/ionization, a pulsed carbon dioxide laser is focussed to approximately a 1 mm diameter spot *via* a ZnSe lens onto a steel circular probe tip (10 mm diameter) attached to a direct insertion probe holding the sample. Following the laser pulse, an appropriate delay (usually 3-5 seconds) is introduced to allow the high local pressure ($10^{-7}$-$10^{-6}$ torr uncorrected ionization gauge reading) of desorbed neutrals to be pumped away, leaving the ions which are trapped to be detected with improved resolution. Such delays should be substantially reduced in the newer differentially pumped dual cell design. Subsequently, the ions are excited to allow their image currents to be recorded using a high speed analog to digital (A/D) converter. The transient signal thus obtained is then apodized and Fourier transformed to produce the mass spectrum for the ions resulting from a single laser pulse. Although signal averaging may be employed, the current spectra all result from single laser pulses using ~1-5 μg of sample. No definitive comparisons of the effects of lasers operating at different wavelengths for laser desorption have been reported.

For photodissociation experiments, a special probe was constructed to replace the 12.7 mm diameter direct insertion probe normally employed. It consists of a hollow stainless steel tube which has a 38 mm focal length quartz lens vacuum sealed at one end and an extended hollow probe tip at the other. The beam of a Lambda Physik excimer laser operating at 308 nm was passed through the probe and lens into the FTMS cell through a small hole in the center of the trap plate as shown in Figure 1.

## Sample Preparation

50-100 μg of each peptide was deposited onto the stainless steel probe tip from a methanol/acetic acid solution (1 drop acetic acid in 1 ml of methanol) resulting in sample consumption of ~1 μg per laser pulse based on a 1 $mm^2$ laser spot and uniform sample coverage. Similar amounts of the other samples were applied from a methanol solution. Sensitivity is very dependent upon sample composition and the results presented are typical for these classes of compounds. Poly(phenylalanine) was insoluble and was deposited from a methanol suspension produced by sonication. Approximately 500 µg of poly(phenylalanine) was required to insure adequate coverage of the probe tip, due to the irregular surface produced as a result of evaporation of the solvent.

## Results and Discussion

Peptides. One of the applications that has benefited most from the development of soft ionization methods for mass spectrometry has been the analysis of polar biological materials. Peptides, in particular, have been extensively studied by a variety of soft ionization methods, primarily FAB, with excellent results being obtained for peptides in the 1000-10000 dalton range. LD-FTMS also has proved to be very successful for analysis of simple peptides and an examination of these results should help delineate the analytical potential of LD-FTMS in this important area.

The LD-FTMS spectrum of the decapeptide Angiotensin I is shown in Figure 2 and is a good example of typical results for peptides of this size. An abundant protonated molecular ion $(M+H)^+$ is observed, as well as the N-terminus sequence ions $(Z_9)^+$ to $(Z_7)^+$ with decreasing relative intensity for the smaller fragments; there are no abundant fragment ions with m/z lower than 900. Although a low abundance C-terminus sequence ion $(A_9)^+$ is observed, the N-terminus ions usually predominate, in addition to fragments resulting from losses of neutrals such as $H_2O$, $NH_3$ and $CO_2$. This is in contrast with FAB results in which C-terminus ions are usually observed (11).

As molecular weight is increased, less fragmentation indicative of the peptide's sequence is observed, as can be seen in the spectrum of the undecapeptide, Substance P, in Figure 3. Here, the protonated molecular ion dominates the spectrum with the sodium cationized species $(M+Na)^+$ also observed. Fragment ions are limited to the first sequence ion from the N-terminus as well those arising from losses of $H_2O$ and $NH_3$. $(Z_9)^+$ and $(A_9)^+$ ions at m/z = 1094 and m/z = 1087 are also observed in low relative abundance, as is an ion which may arise from a minor impurity at m/z = 1104. This reduction in useful fragmentation at higher masses is also observed in FAB-MS. Use of higher laser power density increases structurally relevant fragmentation at the expense of the molecular ion and results in lower signal to noise ratios (*i.e.* decreased sensitivity). The effects of laser power on porphyrin fragmentation also have been reported (7).

The LD spectrum of the tridecapeptide neurotensin (Figure 4), shows several differences. Here fragmentation results almost exclusively from small neutral losses. A protonated molecular ion is still one of the major ions observed (m/z = 1673) and some cationized $(K^+)$ fragment species are apparent as well as a small amount of cationized molecular ion (<5%). Resolution has begun to be degraded to some extent, partially due to the decrease in overall ion signal which necessitates transformation of fewer data points to maximize signal to noise ratio.

For the tetradecapeptide (Figure 5), renin substrate, overall ion signal is lower and the protonated molecular ion is less than 50% relative abundance. Extensive fragmentation is observed in

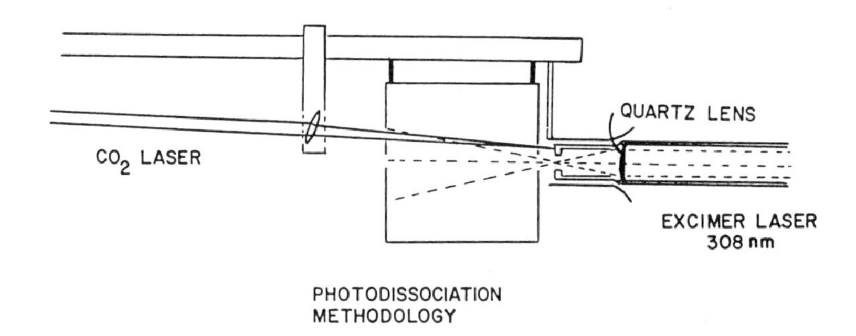


Figure 1. Diagram of the experimental configuration employed for photodissociation experiments.

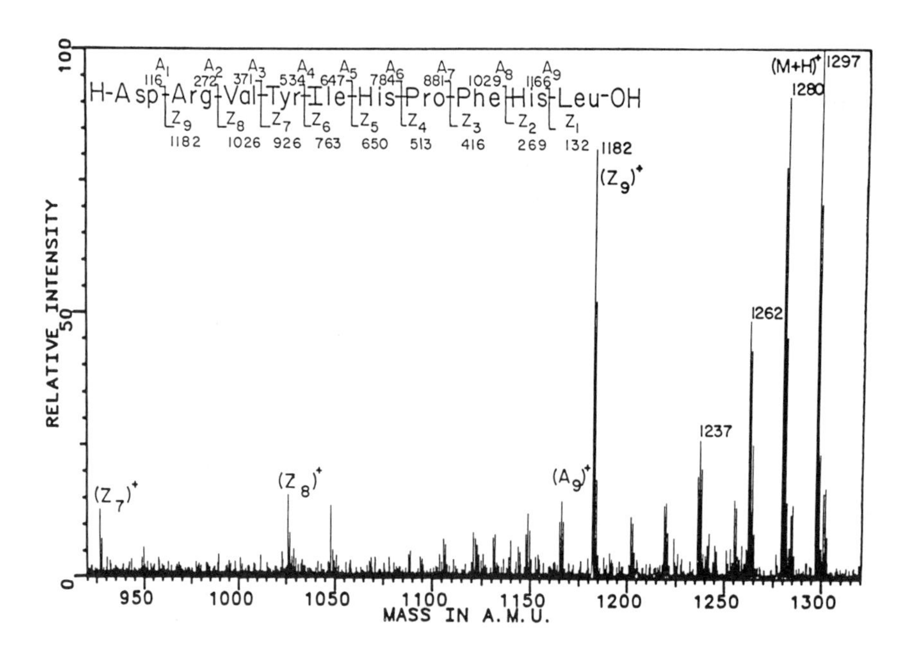


Figure 2. Positive ion spectrum of Angiotensin I.

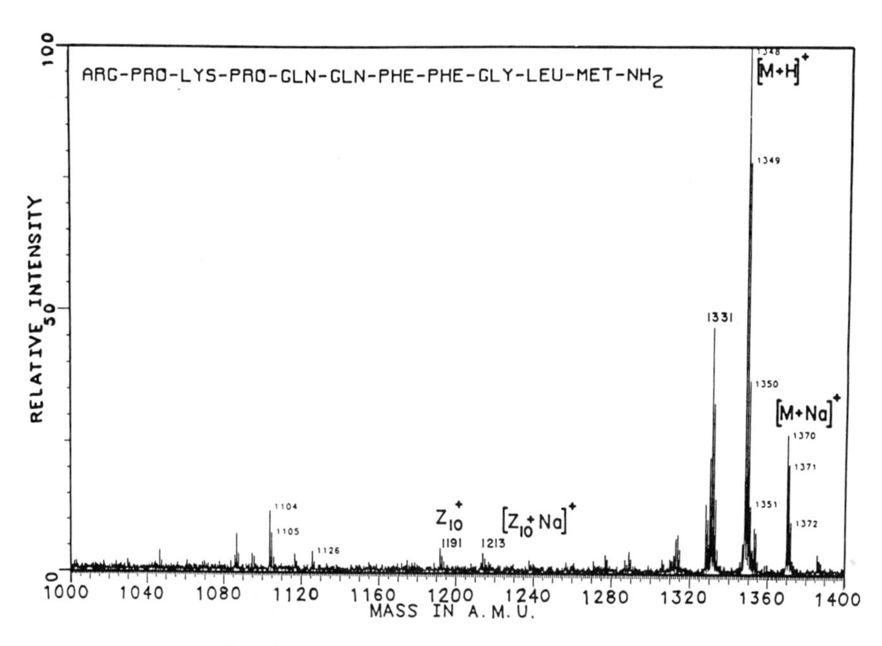


Figure 3. Positive ion spectrum of Substance P.

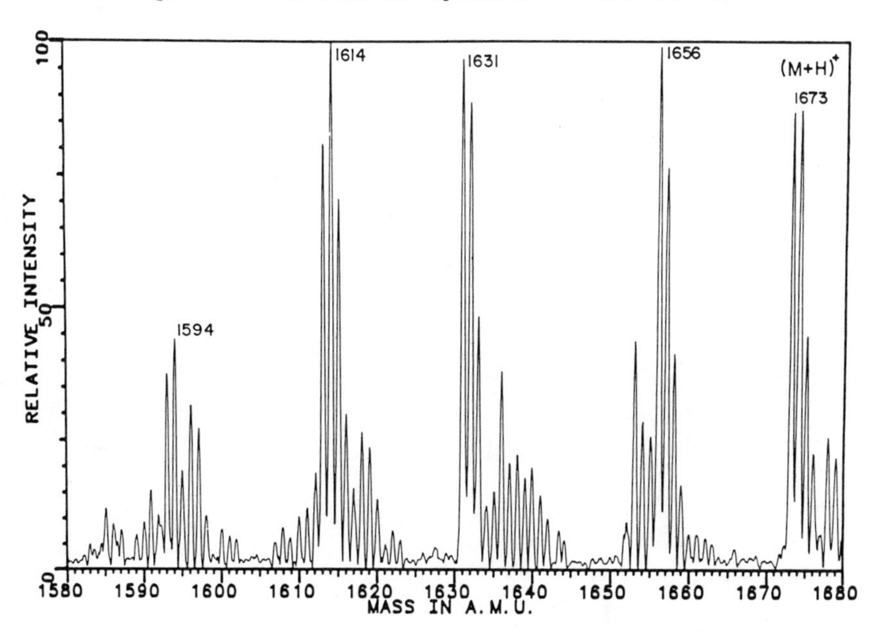


Figure 4. Molecular ion region of the positive ion spectrum of Neurotensin.

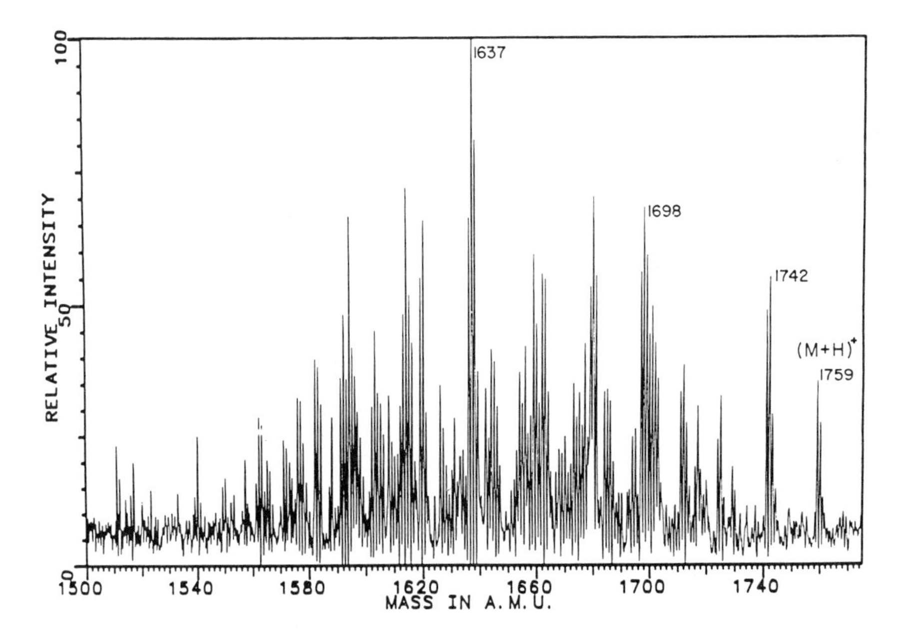


Figure 5. Molecular ion region of the positive ion spectrum of Renin Substrate.

the molecular ion region due to complicated losses of small neutrals. Although some structurally important ions may be present, they are difficult to identify due to the very extensive and complex fragmentation pattern. This extensive fragmentation may be peculiar to renin substrate, as it has not been observed for other peptides.

To date, renin substrate is the second highest molecular weight peptide successfully run by LD-FTMS. The highest mass oligopeptide LD-FTMS spectrum yet obtained is that of Gramicidin A (M.W. 1881) (12) doped with KBr. LD-FTMS of Gramicidin A on our system produces predominately $(M+K)^+$ ions as well as lower intensity N-terminus sequence ions extending from $Z_{14}$ to $Z_5$. This peptide is a typical and not indicative of the behavior expected from most oligopeptides. Further study is planned to try to determine why Gramicidin A is so much easier than similar molecular weight peptides to ionize by LD-FTMS. A representative list of peptides which have been successfully analyzed by LD-FTMS in our laboratory is shown in Table I.

Studies of polymers (2,5) with masses greater than 7000 dalton show the mass range of FTMS is sufficient to accomodate higher mass substances. From our attempts to examine peptides in the 2000-3000 dalton range such as mellitin, chain A insulin, and chain B insulin it appears that some barrier is reached around mass 2000 for peptides. Coincidentally, this is where secondary structure of peptides begins to become important, as the peptide chains can now unfold in solution, which may make it harder to ionize intact molecular ions. Attempts at modifying the surface-peptide interaction with glutathione (13) and nitrocellulose (14) as is done in $^{252}Cf$ plasma desorption have met with no success. Further study is obviously needed.

Table I. Representative Peptides Successfully Examined by LD-FTMS

| Peptide | Molecular Formula | Molecular Weight |
|---|---|---|
| Angiotensin II | $C_{50}H_{89}N_{13}O_{12}$ | 1045.5 |
| Bradykinin | $C_{50}H_{73}N_{15}O_{11}$ | 1059.6 |
| Gramicidin S | $C_{60}H_{92}N_{12}O_{10}$ | 1140.7 |
| Angiotensin I | $C_{62}H_{89}N_{17}O_{14}$ | 1295.7 |
| Substance P | $C_{63}H_{98}N_{18}O_{13}S$ | 1346.7 |
| Bacitracin A | $C_{66}H_{103}N_{17}O_{16}S$ | 1421.7 |
| Bombesin[a] | $C_{71}H_{110}N_{24}O_{18}S$ | 1618.8 |
| Somatostatin | $C_{76}H_{104}N_{18}O_{19}S_2$ | 1636.7 |
| α-melanocyte stimulating hormone | $C_{77}H_{109}N_{21}O_{19}S$ | 1663.8 |
| Neurotensin | $C_{78}H_{121}N_{21}O_{20}$ | 1671.9 |
| Renin Substrate[b] | $C_{85}H_{123}N_{21}O_{20}$ | 1757.9 |
| Gramicidin A | $C_{99}H_{140}N_{20}O_{17}$ | 1881.1 |

[a] $(M+H)^+$ of low abundance
[b] Extensive Fragmentation

Polymers

To model higher molecular weight peptides, the inexpensive biopolymer poly(phenylalanine) (with average molecular weight 2000) was examined. Although this polymer consists of only phenylalanine repeating units, it should be a good model for oligopeptides of similar molecular weights with the added complication that it is quite insoluble in most solvents, with the exception of some concentrated acids. A spectrum of poly(phenylalanine) is shown in Figure 6. Although it exhibits approximately the molecular weight distribution expected, reproducibility from sample to sample was poor; often no ions or only lower mass ions were observed. While this behavior is not unique to this sample, it is unusual in the sense that most samples either do or do not produce particular ions. Reproducibility problems generally arise more often with spectral S/N or resolution fluctuations for a given set of experimental conditions. In addition, the signal was severely damped in all cases, causing extreme resolution degradation. The exact cause of this signal dampening is not yet known and does not affect all higher mass samples (see below). However, the spectrum of this sample with ions approaching m/z 4000 suggests there is hope for increasing the upper mass limit for peptides, as well as other difficult to ionize samples. The distribution is characterized by two major repeating sequences differing by the loss of $H_2O$ with each successive ion 147 amu and lower, corresponding to a to a phenylalanine residue. However, the higher mass ion in each pair does not correspond to any simple fragmentation, nor are these ions present in the spectrum of tetraphenylalanine, which produces an abundant molecular ion [$(M+H)^+$,$(M+K)^+$]. It is probable that a different end group is present due to the introduction of a species to control molecular weight by terminating the polymerization process. This is suggested both by the fragmentation behavior of the lower molecular weight tetraphenylalanine as well as the agreement of the observed molecular weight distribution with the manufacturer's value, indicating little fragmentation.

An example of the resolution which can currently be obtained for a well-behaved sample, is the spectrum (Figure 7) of poly(ethyleneglycol) (PEG) of average molecular weight 3350, showing unit mass resolution to almost 4000 amu. This is the current upper mass limit for unit mass resolution with our 3 Tesla system. We are hoping to improve upon this with a recently-installed 7.2 Tesla system. The resolution for this sample is approximately 7500 at mass 3000. A lower molecular weight PEG of average molecular weight 1450 was mixed with the PEG 3350 and the broadband (100 KHz) resolution of 15000 at mass 1500 obtained extrapolates linearly to higher mass as can be seen in Figure 8. Although some scatter is apparent, it appears that for these samples, the predicted linear loss in resolution with increasingly mass is obeyed.

Photodissociation

As the series of peptide spectra already presented has shown,

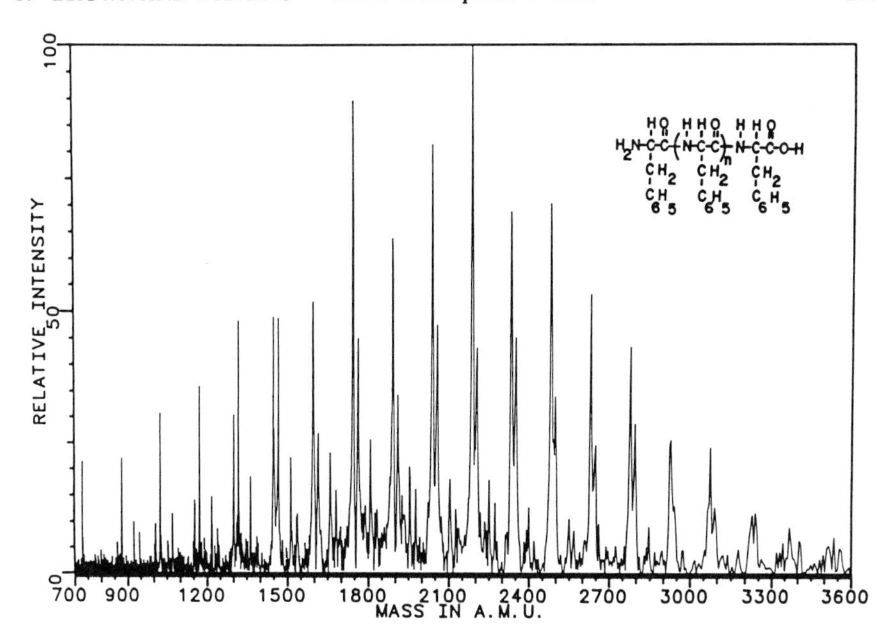


Figure 6. Positive ion spectrum of poly(phenylalanine), average molecular weight 2000.

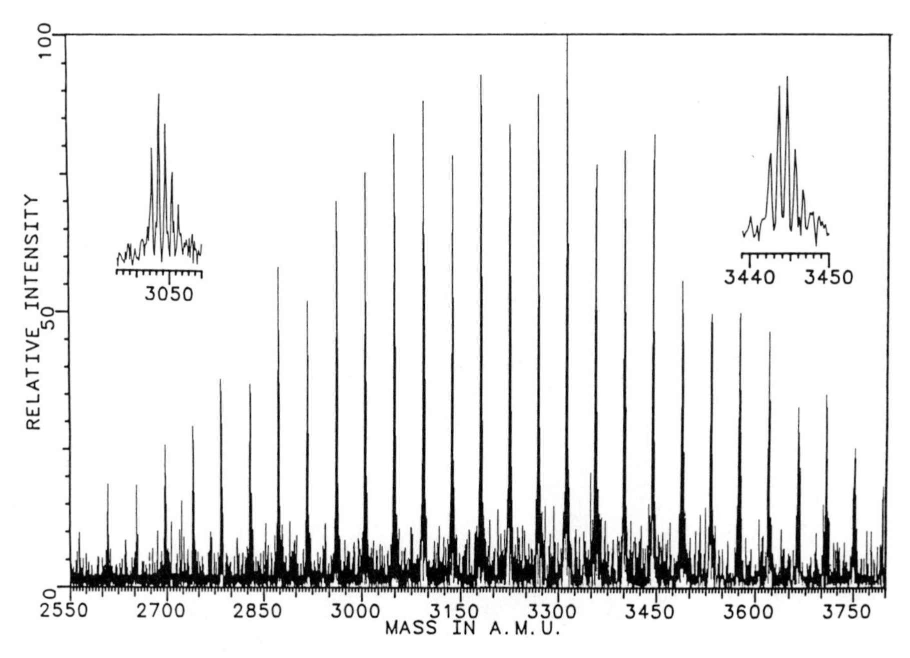


Figure 7. High mass region of the positive ion spectrum of a mixture of (polyethyleneglycol) of average molecular weight 1450 and 3350.

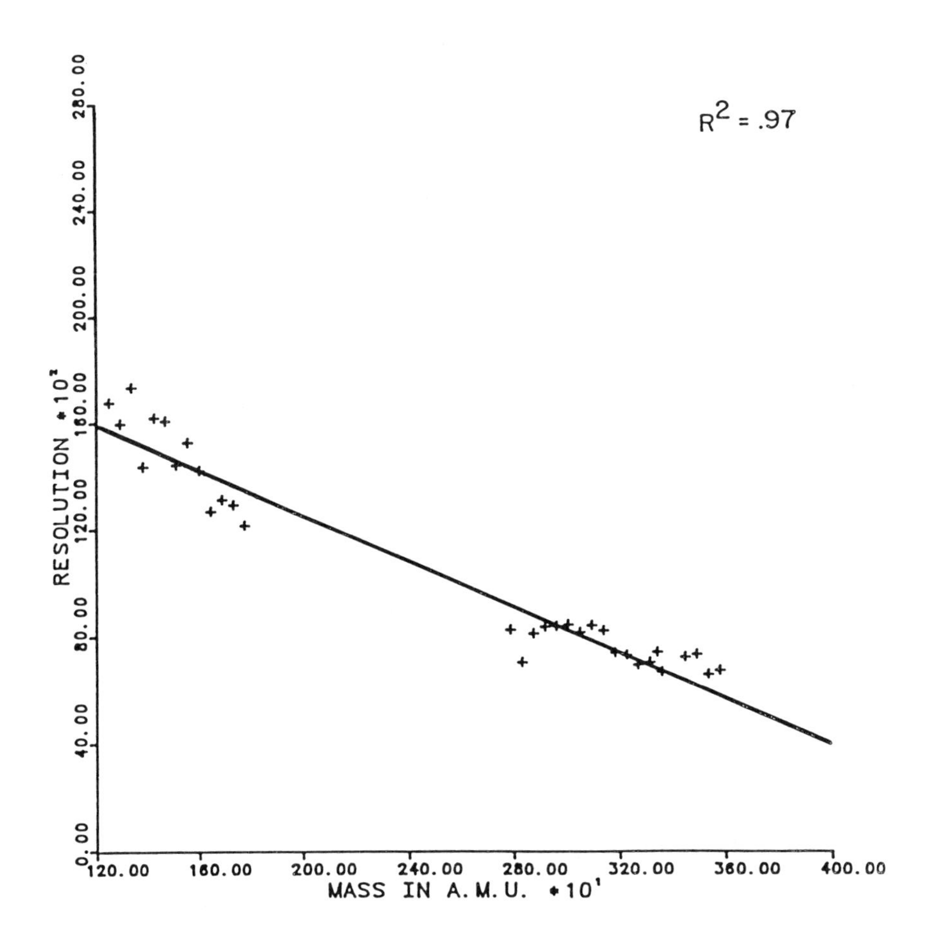


Figure 8. Plot of resolution vs. mass for mixture producing the spectrum in Figure 7.

the amount of fragmentation useful for structural determination drops off substantially at higher masses and is not unique to LD-FTMS. The use of ion activation methods to induce further fragmentation should provide more structurally important fragments. Collision-activated dissociation (CAD), which is typically employed, suffers from poor efficiency at higher masses. An alternative approach employing photodissociation is especially well suited to FTMS. By employing two lasers, one to desorb and ionize (a $CO_2$ laser) followed by a second laser pulse (an excimer laser) to photodissociate the trapped ions, a significant improvement in structurally relevant fragmentation should be possible. Work by McIver (15) has already demonstrated photodissociation of simple peptide ions which had been formed by chemical ionization methods. By using a pulsed $CO_2$ laser to generate the ions, more non-volatile samples may be explored. In addition, multiphoton ionization studies of the desorbed neutrals produced by the pulsed $CO_2$ laser should also be possible.

Our excimer laser is currently used to pump a dye laser at 308 nm. Although this is not an optimum wavelength for photodissociation of most molecules, a preliminary result of the two laser experiment is shown in Figure 9. Figure 9a is the pulsed $CO_2$ laser desorption spectrum of Rhodamine B, with all ions below the $(M-Cl)^+$ ion ejected. Figure 9b results from using an identical pulse sequence with the excimer fired after the normal pump-down delay (5 sec.) for laser desorbed neutrals to be removed from the cell. The spectra are plotted on the same absolute intensity scale. The ions produced appear to result from successive losses of small neutral species, with the major fragment being loss of (HCOOH) from the $(M-Cl)^+$ or loss of (NaCOOH) or (KCOOH) from the cationized species. Subsequent additional losses of $(C_2H_4)$ and $[(C_2H_4)N(CH_2CH_5)]$ account for the majority of ions. It should be noted that all of these ions are also produced in the laser desorption spectrum prior to ejection, but in much lower abundances. Although this is not a particularly difficult sample, these preliminary results are encouraging. Future plans include use of a lower UV wavelength, which should be applicable to a wider range of samples. We also intend to explore the possibility of examining the desorbed neutrals by multiphoton ionization. Accurate mass measurements on the fragments, which were not performed for the current work, should help in assigning fragment composition as well as understanding fragmentation pathways. Accuracies of ~1 ppm error for ions produced by LD-FTMS with masses of over 1300 amu have been reported (7). The usefulness of such data for high mass ions (>1000 amu) is questionable due to the preponderance of molecular compositions at these masses which are within even a 1 ppm error.

## Conclusions

The analytical applications of LD-FTMS have just begun to be developed. As more work is done, LD-FTMS will continue to be applied to analysis of difficult samples. It is a promising technique for stable samples such as polymers, organometallics and

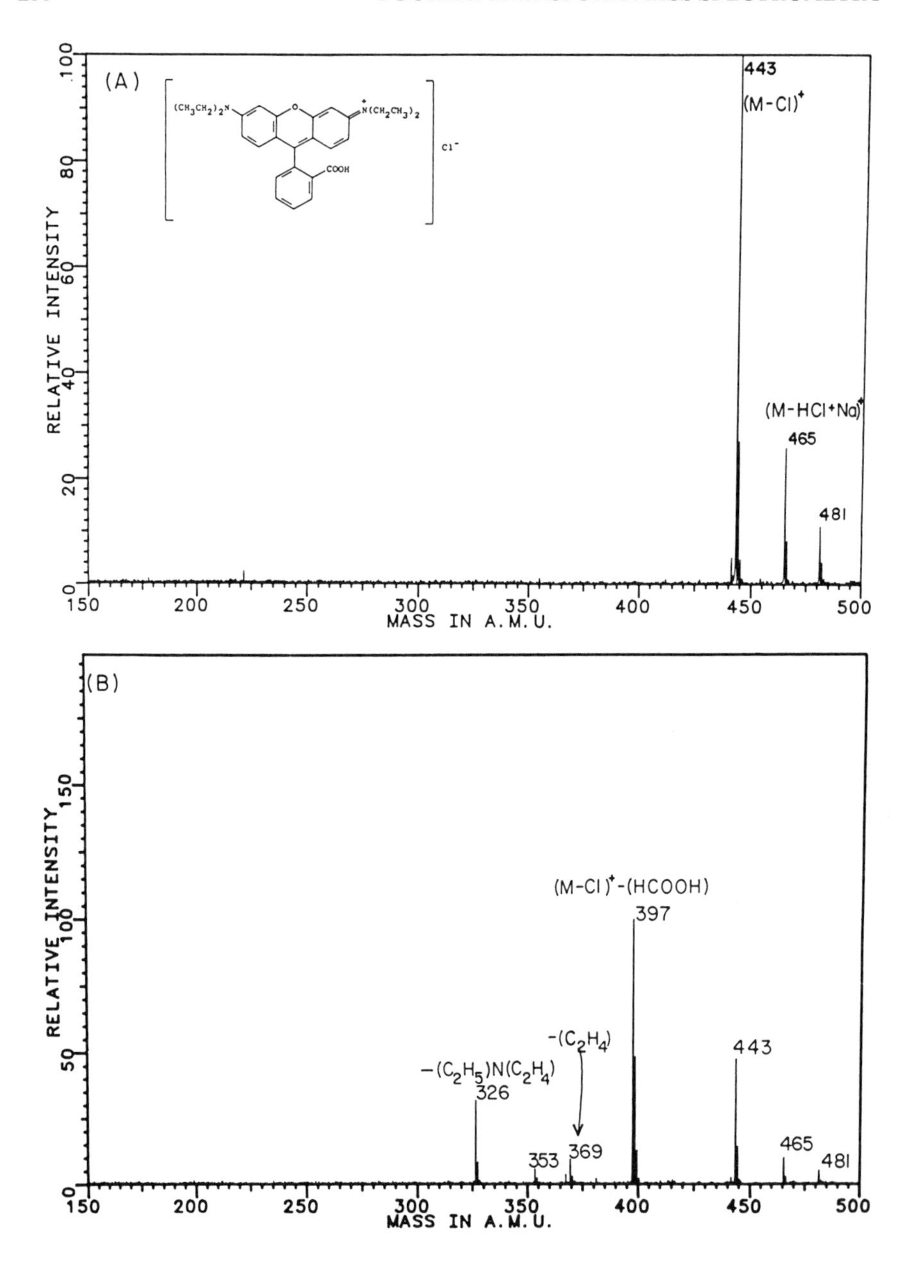


Figure 9. (a) Laser desorption positive ion spectrum of Rhodamine B utilizing only the $CO_2$ laser with ejection of all ions below m/z = 443.
(b) Positive ion spectrum as in (a) but with UV irradiation (308 nm) after the pump down delay.

porphyrins, as well as many insoluble samples. Continued work in the area is expected to improve both the mass range and sensitivity while better instrumentation and understanding of the ion dynamics of FTMS should result in the realization of the high resolution promise of FTMS at high mass. Given its present state of development, LD-FTMS can provide unique capabilities for solution of a variety of analytical problems.

## Acknowledgments

Support from the National Institutes of Health under grant GM-30604 is gratefully acknowledged.

## Literature Cited

1. McCrery, D.A.; Ledford, E.G., Jr.; Gross, M.L. Anal. Chem. 1982, 54, 1435-1437.
2. Wilkins, C.L.; Weil, D.A.; Yang, C.L.C.; Ijames, C.F. Anal. Chem. 1985, 57, 520-524.
3. Wilkins, C.L.; Yang, C.L.C. Int. J. Mass Spectrom. Ion Processes 1986, 72, 195-208.
4. McCrery, D.A.; Gross, M.L. Anal. Chim. Acta, 1985, 178, 103-116.
5. Brown, R.S.; Weil, D.A.; Wilkins, C.L. Macromolecules 1986, 19, 1255-1260.
6. Brown, C.E.; Kovacic, P.; Wilke, C.E.; Cody, R.B.; Kinsinger, J.A. J. Polymer Sci. Poly. Lett. Ed., 1985, 23, 453-463.
7. Brown, R.S.; Wilkins, C.L. Anal. Chem. 1986, 58, 3196-3199.
8. Brown, R.S.; Wilkins, C.L. J. Am. Chem. Soc. 1986, 108, 2447-2448.
9. Coates, M.L.; Wilkins, C.L. Biomed. Mass Spectrom. 1985, 12, 424-428.
10. McCrery, D.A.; Gross, M.L. Anal. Chim. Acta, 1985, 178, 91-103.
11. König, W.A.; Aydin, M.; Schulze, V.; Rapp, V.; Hohn, M.; Pesch, R.; Kalikhevitch, V.N. Int. J. Mass Spectrom. Ion Phys. 1983, 46, 403-406.
12. Cody, R.B., Jr.; Amster, I.J.; McLafferty, F.W. Proc. Natl. Acad. Sci. USA 1985, 82, 6367-6370.
13. Alai, M.; Demirev, P.; Fenselau, C.; Cotter, R. Anal. Chem. 1986, 58, 1303-1307.
14. Jonsson, G.P.; Hedin, A.B.; Hakansson, P.L.; Sundquist, B.U.; Save, G.B.S.; Nielsen, P.F.; Roepstorff, P.; Johansson, K.E.; Kamensky, I.; Lindberg, M.S.L. Anal. Chem. 1986, 58, 1084-1087.
15. Bowers, W.D.; Delbert, S.S.; McIver, R.T. Anal. Chem. 1986, 58, 969-972.

RECEIVED September 14, 1987

Chapter 9

# Infrared Multiphoton Dissociation of Laser-Desorbed Ions

Clifford H. Watson, Gökhan Baykut [1], and John R. Eyler

Department of Chemistry, University of Florida, Gainesville, FL 32611

Output from both gated continuous wave and pulsed carbon dioxide lasers has been used to desorb ions from surfaces and then to photodissociate them in a Fourier transform ion cyclotron resonance mass spectrometer. Pulsed $CO_2$ laser irradiation was most successful in laser desorption experiments, while a gated continuous wave laser was used for a majority of the successful infrared multiphoton dissociation studies. Fragmentation of ions with m/z values in the range of 400-1500 daltons was induced by infrared multiphoton dissociation. Such photodissociation was successfully coupled with laser desorption for several different classes of compounds. Either two sequential pulses from a pulsed carbon dioxide laser (one for desorption and one for dissociation), or one desorption pulse followed by gated continuous wave irradiation to bring about dissociation was used.

The technique of laser desorption (LD) has been widely used in mass spectrometry (1,2) to desorb and to ionize high molecular weight or other nonvolatile samples, most often using time-of-flight (3,4) or Fourier transform ion cyclotron resonance (FTICR) (5,6) mass spectrometers for mass analysis. In this technique, highly focussed laser irradiation, most often with a power density of at least $10^8$ W/cm$^2$, is used to desorb and ionize a solid sample that has been inserted into the high vacuum system of the mass spectrometer.

The primary advantage of laser desorption is that abundant molecular or pseudomolecular ions are produced for many different classes of compounds. Positive pseudomolecular ions are most often formed by attachment of a cation, typically a proton, potassium, or sodium ion, to the parent molecule. Negative pseudomolecular ions can also be formed by laser desorption, usually by loss of a proton, or by electron attachment to molecules with a positive electron affinity. Often little fragmentation occurs (making this

[1]Current address: Ion Spec, One Longstreet, Irvine, CA 92714

0097-6156/87/0359-0140$06.00/0

one of the "soft" ionization techniques), and the strong molecular ion signal ($M^+$, $(M + H)^+$, $(M + Na)^+$, $(M + K)^+$, $M^{-\bullet}$, or $(M - H)^-$) provides molecular weight information.

Although an abundant molecular ion peak is important in determination of the molecular weight of a compound, fragmentation is often desirable to characterize the molecular structure. Several techniques have been applied to achieve fragmentation; most notable are collision induced dissociation (7) and photodissociation (8,9). Thus, it would be advantageous to combine one of these techniques with laser desorption to obtain both molecular weight and structural information. One possible drawback of collision induced dissociation is difficulty in dissociating larger ions (10-12) because of an inability to impart sufficient internal energy to them in collisions with much lighter target gases.

Fourier transform ICR mass spectrometry is ideally suited for LD ionization because, as with time-of-flight (TOF) mass spectrometers, a complete mass spectrum can be obtained for every laser pulse. In addition, FTICR analysis offers much higher mass resolution than can be attained by TOF, even after modifications to the TOF instrument such as the addition of reflecting grids. High mass resolution can be quite important in the desorption and fragmentation of large molecules, including those of biological relevance.

Earlier studies from our laboratory (13,14) have shown the feasibility of coupling gated continuous wave (cw) $CO_2$ lasers with an FTICR mass spectrometer to photodissociate ions trapped there. The ability to hold ions in a constrained irradiated volume permits facile photodissociation of ions with low to moderate molecular weights. Combination of these two laser techniques (LD to produce ions and photodissociation (PD) to fragment them) was thus of great interest. This chapter discusses the success of this coupling when applied to compounds with molecular weights in the range of 100-600 daltons.

## Experimental Section

Experiments were carried out in a Nicolet FT/MS - 1000 mass spectrometer. The standard laser desorption interface supplied by the manufacturer was used. As shown by the solid line in Figure 1, light from a Lumonics TE 860 grating-tuned pulsed $CO_2$ (infrared) laser entered the vacuum chamber through a ZnSe window and was focused by a ZnSe lens of 1.25 cm diameter and 5 cm focal length onto a solids probe containing the analyte. An Apollo 570 continuous wave grating-tuned $CO_2$ (infrared) laser, gated by a trigger pulse from the FTICR electronics, was also employed for laser desorption. However, its output was of insufficient power density to generate ions reliably and so its usage was limited.

Both the pulsed and cw lasers were used for photodissociation of gaseous ions produced by laser desorption. The cw laser had a maximum output power of 50 watts at 10.61 micrometers and a beam diameter of 6 mm. The pulsed laser produced 2.6 joules in a pulse of 1 microsecond duration at 10.61 micrometers and had a 2 x 3 cm rectangular beam shape. Modifications of the FTICR vacuum chamber that facilitate ion irradiation have been reported previously (13).

In the "one laser experiment" (Figure 1), the ions were desorbed by a single laser pulse using a focussing lens. A mirror outside the vacuum chamber was then rotated so that a second unfocussed laser pulse entered the analyzer cell and irradiated the stored ions. The external mirror was then returned to its original position for the next experiment. Unfortunately, a very large $K^+$ signal was observed for both the focussed and unfocussed pulses, and the unfocussed laser pulse often desorbed contaminant neutrals from the cell plate opposite the beam reflector shown in Figure 1.

Difficulty in eliminating $K^+$ ions and contaminant neutrals formed during the second laser pulse, and problems with alignment of the laser beam when moving the external mirror suggested the use of a second laser, the cw $CO_2$ laser, to photodissociate ions produced by the pulsed laser. In the "two laser" experiment, shown in Figure 2, the conventional Nicolet solids probe was replaced by a NaCl window mounted on a hollow stainless steel tube that extended into the vacuum chamber to within a few millimeters of the FTICR cell. Typically 1-5 mg of analyte was dissolved in 3-5 ml of either acetone or ethanol and a portion of this solution was deposited using a micropipette onto the opposite end of the tube from the window and nearest to the cell. This assembly served as both sample support and as a second window for laser irradiation of the ions. Following sample preparation, the tube assembly was inserted into the vacuum chamber through the solids probe inlet. As shown in Figure 2, the pulsed laser beam was directed into the cell using the same optics as in Figure 1 to desorb the ions in the normal manner. The (gated) cw laser beam entered the vacuum chamber through the salt window mounted on the tube. Ions were formed by LD using focussed light from the pulsed laser. After a delay of 1-3 seconds to allow the initial pressure burst due to desorbed neutrals to dissipate, the cw laser irradiated the ions for 0.5-3 seconds until significant photodissociation was observed.

The hollow tube could be rotated, and its tip was divided into 8 different areas in which there was no overlap of the pulsed laser beam. Samples deposited on each area produced significant ion signal for a range of 3-20 laser pulses, depending on the sample and preparation characteristics. The standard experimental method used involved collection of an LD, an LD/PD, and a second LD mass spectrum. This procedure provided a reference spectrum with no photodissociation both before and after the LD/PD spectrum, and eliminated the possibility of observed photofragments originating from some unknown source.

Photodissociation of larger ions (>200 daltons) formed by electron impact was also studied. These ions were produced by electron ionization of vapor sublimed from the standard heated solids insertion probe. Probe temperatures of 150-200°C were employed. This process gave rise to a low abundance of molecular ions. Therefore a variable length (100-1000 ms) reactive delay time was utilized to produce ions by ion/molecule reactions. Fragment ions formed by electron impact reacted with the neutral background molecules to produce the various ions of interest.

Samples were obtained from commercial sources, or acquired from various research laboratories. Sample purity was confirmed by wide mass range spectra, and the samples were used without further purification.

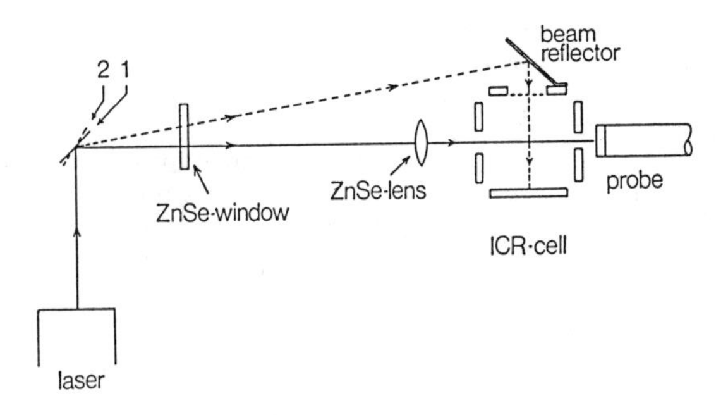


Figure 1. Configuration used for single laser experiments with a pulsed $CO_2$ laser. During the first laser pulse, the beam was focussed through the lens and formed ions by laser desorption. A rotatable mirror directed the second, unfocussed laser beam into the cell to dissociate trapped ions by multiphoton dissociation.
(Reproduced from ref. 18. Copyright 1987 American Chemical Society.)

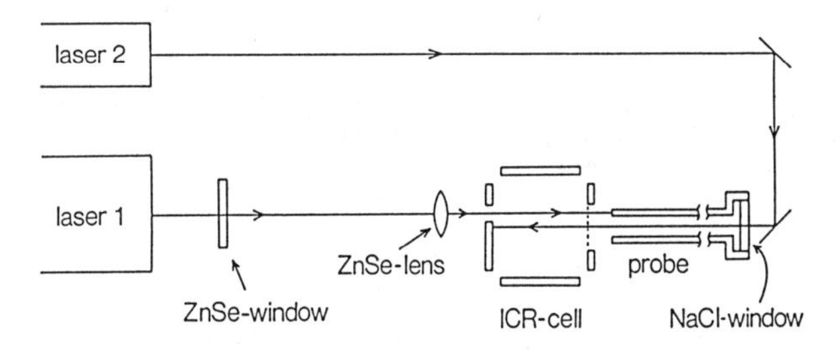


Figure 2. Configuration used for double laser experiment. A pulsed $CO_2$ laser (laser 1) was used to desorb and ionize the sample and a cw $CO_2$ laser (laser 2) was used to dissociate the trapped ions.
(Reproduced from ref. 18. Copyright 1987 American Chemical Society.)

## Results and Discussion

Photodissociation of large ions. Before attempting photodissociation of moderate molecular weight ions formed by laser desorption, initial experiments were carried out to verify that ions in the 500-1500 dalton range of molecular weight could, in fact, be made to dissociate by the absorption of $CO_2$ laser irradiation. These larger ions, with their many degrees of freedom and higher density of states, are expected to appear "black" in the infrared region; that is, their infrared absorption spectrum should be nearly continuous, leading to a high probability that they will absorb infrared photons. The question of whether internal excitation produced by infrared absorption would be relaxed by radiative or collisional processes faster than such energy might accumulate in certain modes leading to bond rupture was of great interest. Three compounds were studied using electron impact ionization to see if earlier work (13,14) on ions of molecular weights in the 50-200 dalton range could in fact be extended to ions of substantially higher m/z values.

An ion of nominal m/z 442 was produced from cis-dichloro-trans-dihydroxo-bis-2-propanamine/platinum (IV) (CHIP) by ion/molecule reactions (IMR) of the major electron impact (EI) fragment ion with the neutral molecule. As shown in Figure 3, this ion dissociated when irradiated by the cw laser for ca. 500 ms to produce two daughter photofragment ions, m/z 366 and m/z 311, by loss of various ligands (Equation 1).

$$m/z\ 441 + nh\nu \rightarrow m/z\ 366 \text{ and } m/z\ 311 \qquad (1)$$

Shown in Figure 4 is the photodissociation of protonated N,N'-bis(4,6-dimethoxysalicylidene)-4-trifluoromethyl-o-phenylenediiminato cobalt (II) (CoSALOPH), which is also produced by ion/molecule reactions of electron impact fragments with the neutral molecule. This ion dissociated by loss of a methyl group (Equation 2).

$$m/z\ 562 + nh\nu \rightarrow m/z\ 547 \qquad (2)$$

The ion produced by electron attachment to tris(perfluoro-n-nonyl)triazine dissociated when irradiated by light from the cw laser as shown in Figure 5 (Equation 3).

$$m/z\ 1485 \xrightarrow{nh\nu} m/z\ 1066 \qquad (3)$$

This is the highest molecular weight compound for which infrared multiphoton dissociation has been observed in our laboratories to date. A similar pathway was indicated by the collision-induced dissociation spectrum of this ion. Limited CID experiments were performed on some of the other laser-desorbed ions discussed here and in general the CID and PD pathways were the same.

Use of a single laser for desorption/dissociation. Photodissociation of small gaseous ions using a pulsed $CO_2$ laser has been reported (15), so an attempt was made both to form ions and to photodissociate them with sequential laser pulses from the same pulsed laser using the irradiation scheme shown in Figure 1. As

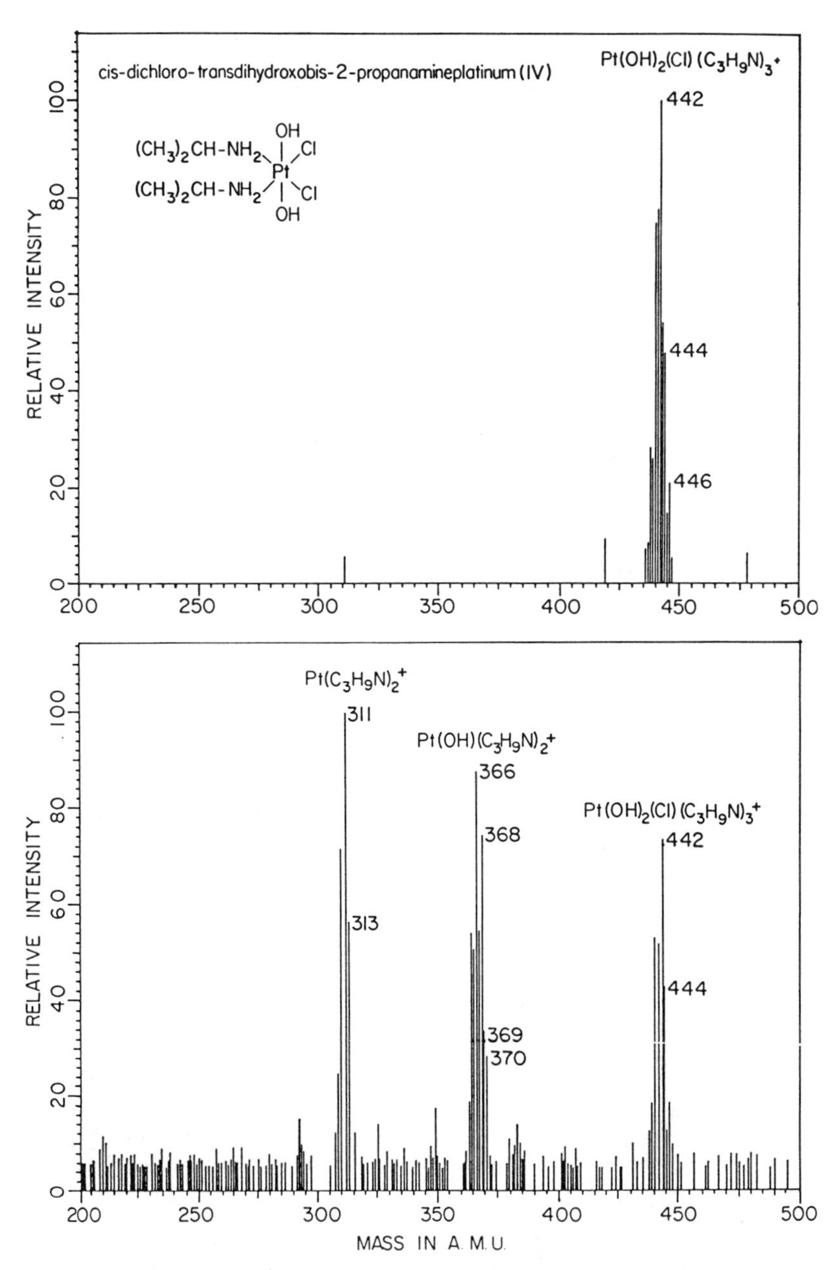

Figure 3. Top. Ion/molecule reaction product formed by reaction of cis-dichloro-trans-dihydroxo-bis-2-propanamine platinum (IV) with its 50 eV electron impact fragments. Bottom. Dissociation pattern produced upon irradiation with cw $CO_2$ laser, same experimental conditions and delay times as in the top spectrum. (Reproduced from ref. 18. Copyright 1987 American Chemical Society.)

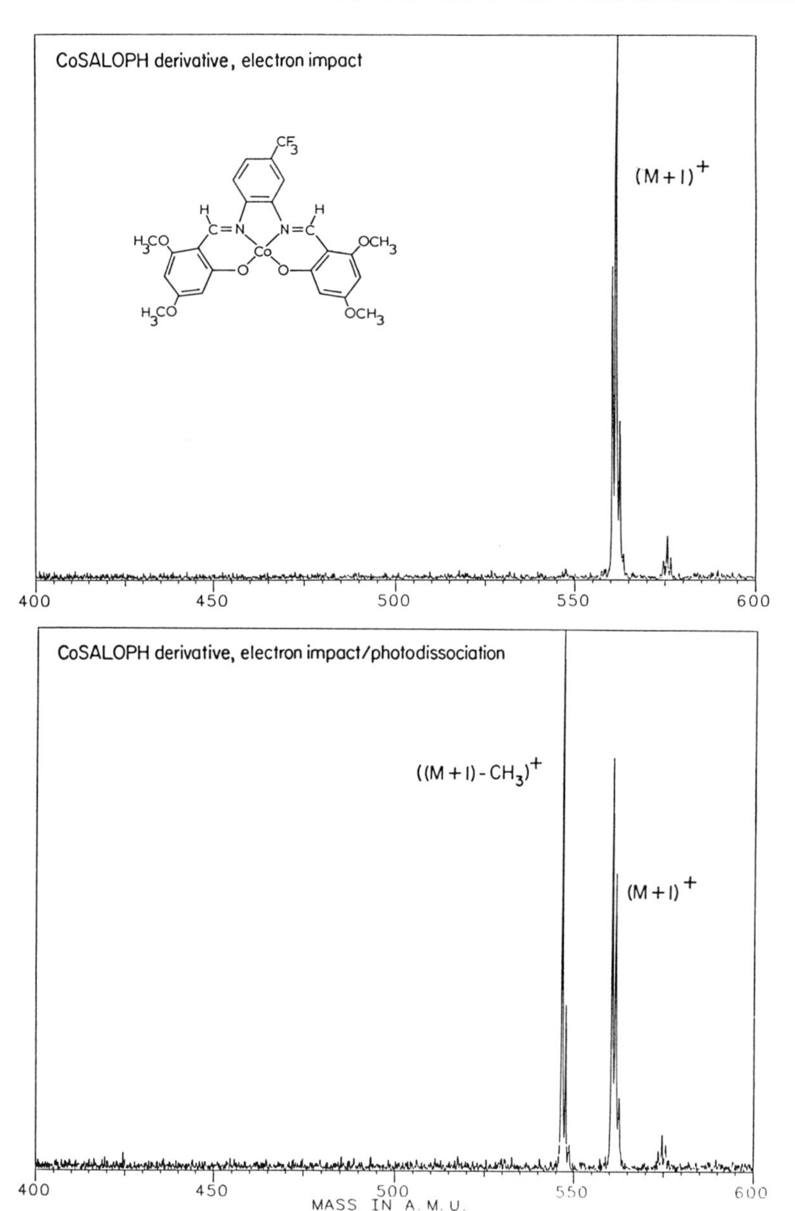


Figure 4. Top. Protonated N,N'-bis(4,6-dimethylsalicylidine)-4-trifluoromethyl-o-phenylenediiminato cobalt (II) (CoSALOPH) formed by reaction of the neutral molecule with fragment ions produced by electron impact at 50 eV. Bottom. Dissociation obtained upon irradiation with the cw $CO_2$ laser, same experimental conditions and delay times as in the top spectrum.
(Reproduced from ref. 18. Copyright 1987 American Chemical Society.)

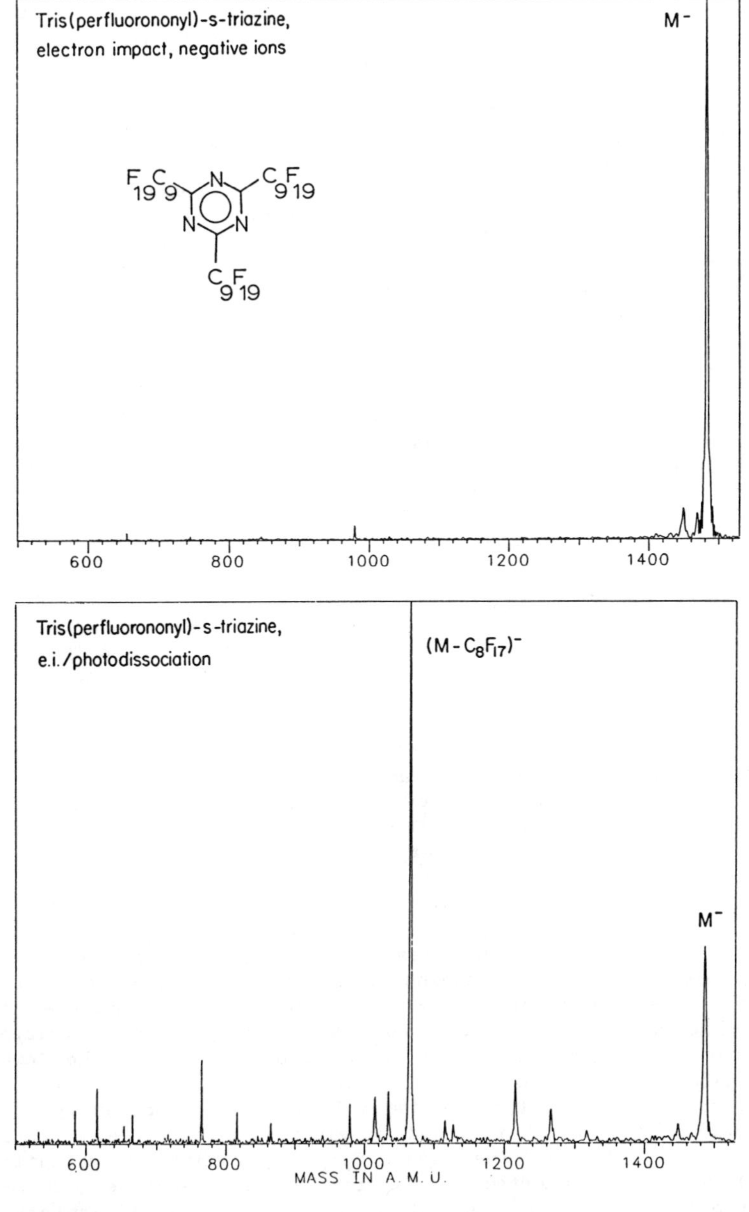


Figure 5. Top. Negative molecular ion of tris-n-nonyl fluorotriazine formed by capture of thermal electrons. Bottom. Dissociation obtained upon irradiation with cw $CO_2$ laser, same experimental conditions and delay times as in the top spectrum. (Reproduced from ref. 18. Copyright 1987 American Chemical Society.)

reported in the experimental section, a large $K^+$ signal always appeared during the pulsed laser desorption experiment. The intensity of the $K^+$ signal was reduced somewhat by adjusting various delay times, and ejecting this ion during the laser beam duration and for 50 ms following the end of the laser pulse. Time resolved LD mass spectrometry (4) has shown that significant amounts of $K^+$ are formed for up to 30 milliseconds following the laser desorption pulse. Under these ejection conditions, photodissociation of the t-butylpyridinium cation (from the perchlorate salt) produced by laser desorption was observed. As shown in Figure 6 the t-butylpyridinium cation, m/z 136, dissociates under pulsed laser irradiation with loss of $C_4H_8$ to produce an ion of m/z 80, which is most likely protonated pyridine (Equation 4).

$$m/z\ 136 + nh\nu \rightarrow m/z\ 80 \qquad (4)$$

The magnitude of the photodissociation effect is quite small in this case. A new FTICR analyzer cell has recently been constructed that will permit passage of the laser beam through the cell without striking any of the cell plates. Gentle focussing of the pulsed laser beam should then increase the laser fluence in the center of the cell, and improve photodissociation efficiency.

Use of two lasers for desorption/dissociation. N,N'-bis(4,6-dimethoxysalicylidene)-4-trifluoromethyl-o-phenylenediiminato cobalt (II) (CoSALOPH): This compound was selected as a test case since its PD pathways were known (see Equation 2 and Figure 4) and significant amounts of $(M + H)^+$ were observed following laser desorption. Photodissociation behavior observed following laser desorption was identical to that seen when electron impact was used to produce the ions (Equation 2 and Figure 4).

The fragmentation observed for this ion is not of great assistance in assigning molecular structure, since only a methyl group is lost from the periphery. This illustrates one potential weakness in the use of infrared multiphoton dissociation (IRMPD) for structure elucidation: the dissociation pathways accessed are those (usually few) of lowest activation energy (9). If these pathways correspond to the loss of small pendant groups from a large molecular ion, no meaningful structural information can be obtained. However, similar behavior would also be expected for these ions with collision-induced dissociation, since inefficient collisional transfer of energy would also only access the lowest activation energy fragmentation pathways.

Sucrose: Observation of the molecular ion produced by laser desorption (16) and LD followed by electron impact (Hein, R. E.; Cody, R. B., Nicolet Instrument Corp., personal communication, 1986.) has been reported for sucrose. The negative ion mass spectrum following laser desorption reveals an $(M - 1)^-$ ion at m/z 341. When irradiated, this ion dissociates with production of ions at m/z 251, 249, 179 and 161 as shown in Figure 7. The ion at m/z 161 corresponds to the loss of a water molecule from one monosaccharide unit. The ion at m/z 179 most likely results from loss of one monosaccharide unit. Since both glucose and fructose have identical molecular formulae, it is not possible at this time to assign a definitive identity to this ion.

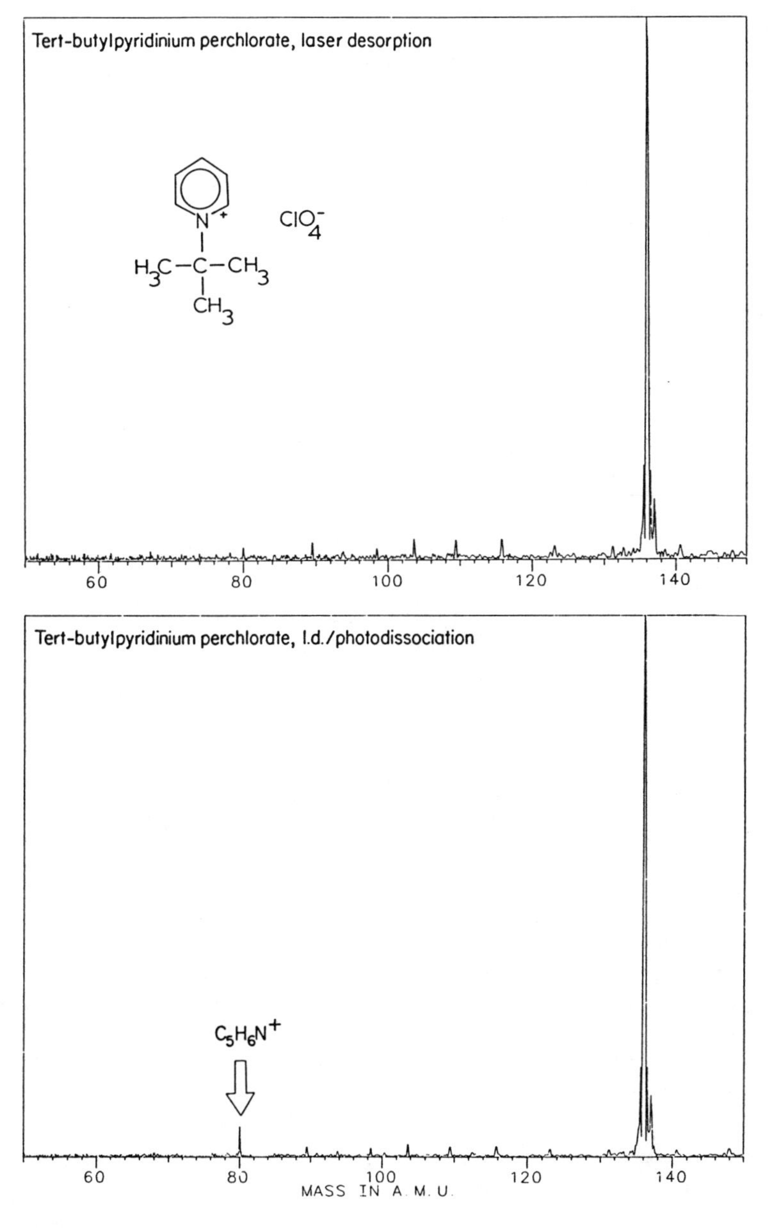


Figure 6. Top. Tert-butylpyridinium cation formed by laser desorption from its perchlorate salt. Bottom. Dissociation achieved with a single laser pulse from a pulsed $CO_2$ laser, all other conditions identical to the top spectrum.
(Reproduced from ref. 18. Copyright 1987 American Chemical Society.)

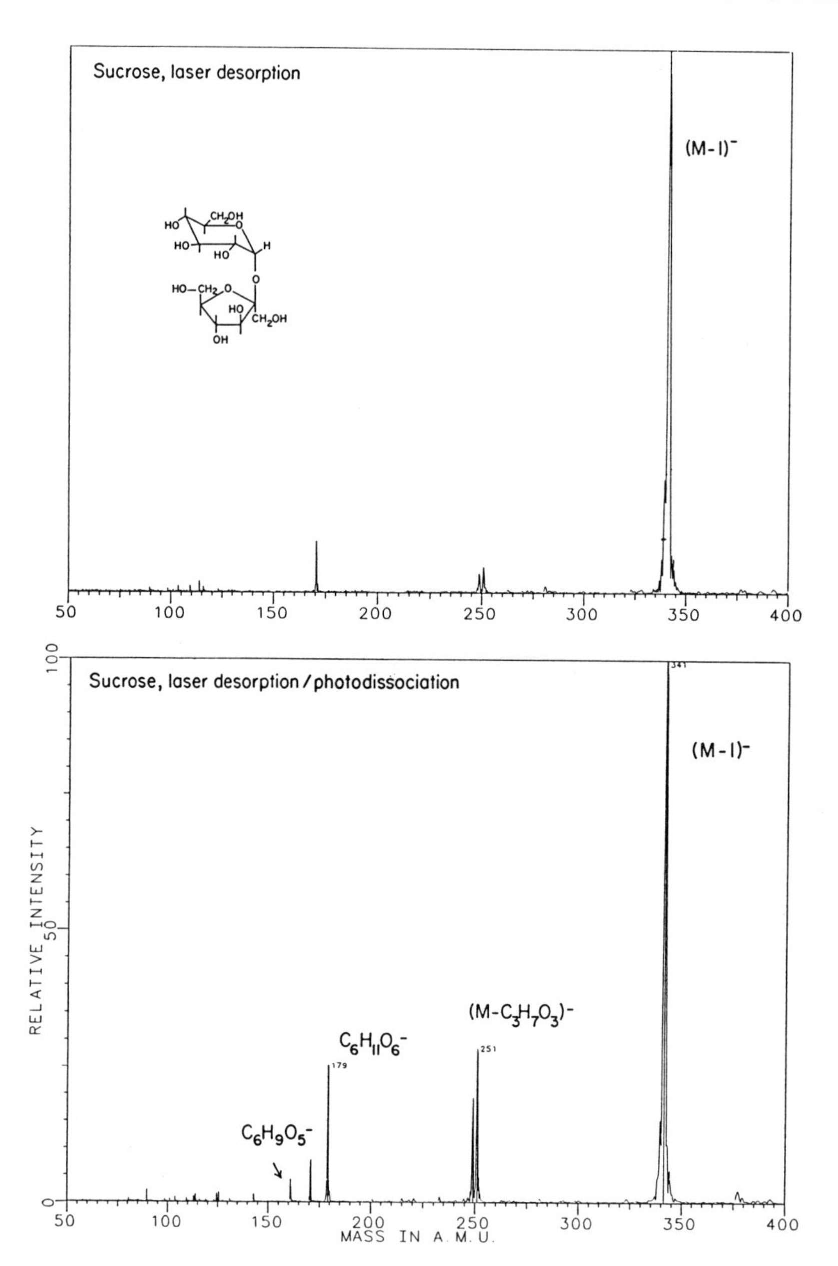


Figure 7. Top. $(M-H)^-$ ion of sucrose formed by desorption with pulsed $CO_2$ laser beam. Bottom. Dissociation upon irradiation with the cw $CO_2$ laser. Experimental conditions are same as in the top spectrum.

The two fragment ions discussed above (m/z 179 and m/z 161) were observed in the negative LD/EI mass spectrum (Hein, R. E.; Cody, R. B., Nicolet Instrument Corp., personal communication, 1986.) of sucrose. The fragment ions with m/z 251 and m/z 249, however, are unique photodissociation products. The ions formed at m/z 251 and 249 correspond to loss of $C_3H_7O_3$, and $C_3H_9O_3$ respectively, suggesting that one of the monosaccharide chains may have opened as a result of ionization; otherwise photodissociation brings about the cleavage of two bonds. It is not known at this time whether these ions are produced by separate pathways from the $(M-1)^-$ ion or from consecutive photodissociations.

Hesperidin: In negative ion LD, an ion of m/z 612 was produced from the sodium salt of hesperidin phosphoric acid ester. This ion was observed to dissociate by loss of the attached sugar to produce an ion at m/z 301 when irradiated by the cw $CO_2$ laser as shown in Figure 8 (Equation 5).

$$m/z\ 612 + nh\nu \rightarrow m/z\ 301 \qquad (5)$$

Recent experiments involving hesperidin itself reveal a similar dissociation as well as production of m/z 327 via a second fragmentation pathway.

N-[2-(5,5-Diphenyl-2,4-imidazolidinedion-3-yl)ethyl]-7-acetoxy-1-napthalene sulfonamide: This substituted sulfonamide has a molecular weight of 543 daltons, and in the positive LD mass spectrum two ions at m/z 566 and m/z 582, corresponding to attachment of sodium and potassium ions, respectively, were observed. No significant photodissociation pathway other than loss of the attached cation was seen for either of these two ions. In the negative ion LD mass spectrum, an ion at m/z 500 originating from loss of $C_2H_3O$ was observed. This ion photodissociated when irradiated by the (gated) cw laser to produce an ion at m/z 251 (see Figure 9) (Equation 6).

$$m/z\ 500 + nh\nu \rightarrow m/z\ 251 \qquad (6)$$

High resolution mass analysis of the ion at m/z 251 indicated that it has the empirical formula $C_{15}H_{11}N_2O_2^-$, and is thus the 5,5-diphenyl-2,4-imidazolidinedione anion, resulting from cleavage of the C-N bond next to the imidazole ring.

## Conclusions

The feasibility of using laser photodissociation to further fragment ions produced by laser desorption has been demonstrated for moderate molecular weight compounds. This technique would be expected to work equally well for both positive and negative ions formed by laser desorption. However, for the compounds reported here, negative ion laser desorption usually produced much higher intensities of molecular or pseudo-molecular ions than the corresponding positive ion mode. Applications of ultraviolet (12) and two photon visible (17) photodissociation have been demonstrated and could be adapted for the photodissociation of laser desorbed ions. We are extending this work to larger molecular weight compounds, and are also modifying our vacuum system to allow

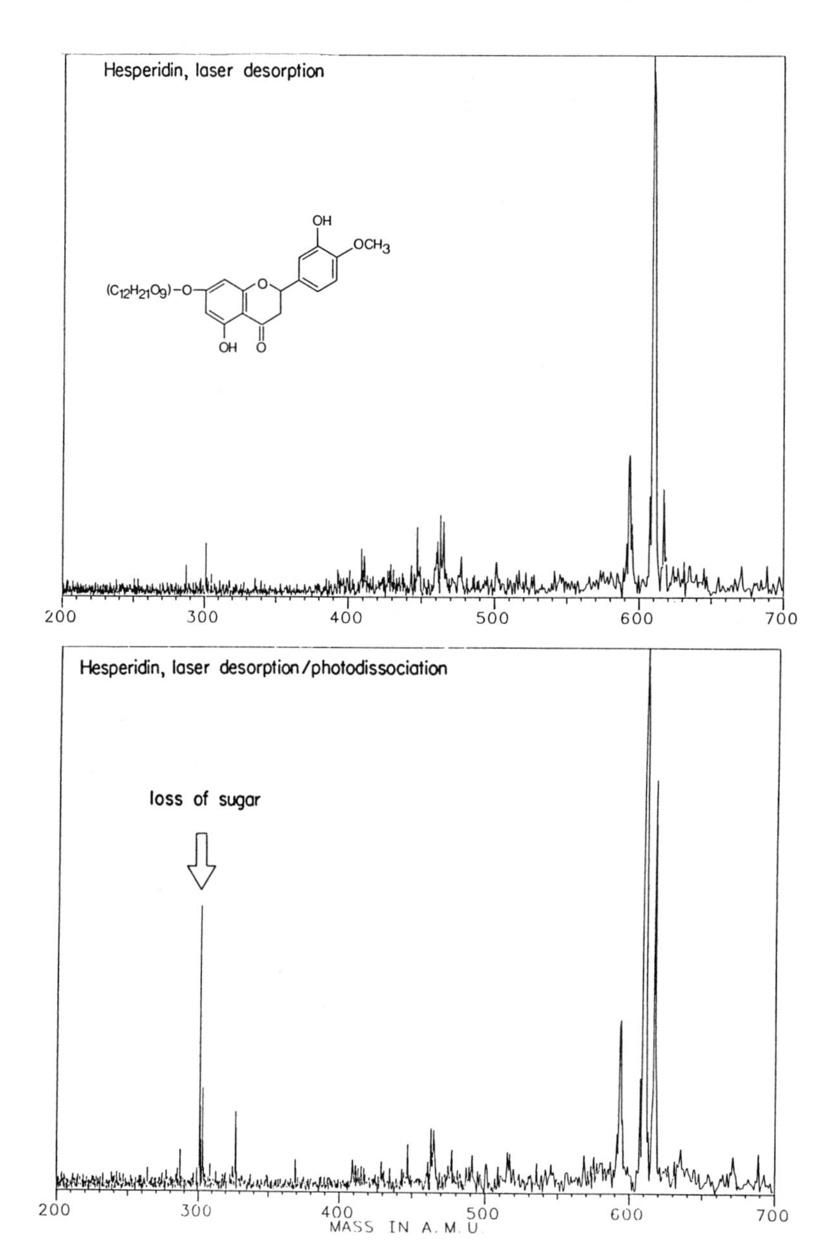


Figure 8. Top. Negative ions formed by laser desorption from Na-salt of hesperidin phosphoric acid ester using pulsed $CO_2$ laser. Bottom. Dissociation upon irradiation with a cw $CO_2$ laser. Experimental conditions are identical to those of the top spectrum. (Reproduced from ref. 18. Copyright 1987 American Chemical Society.)

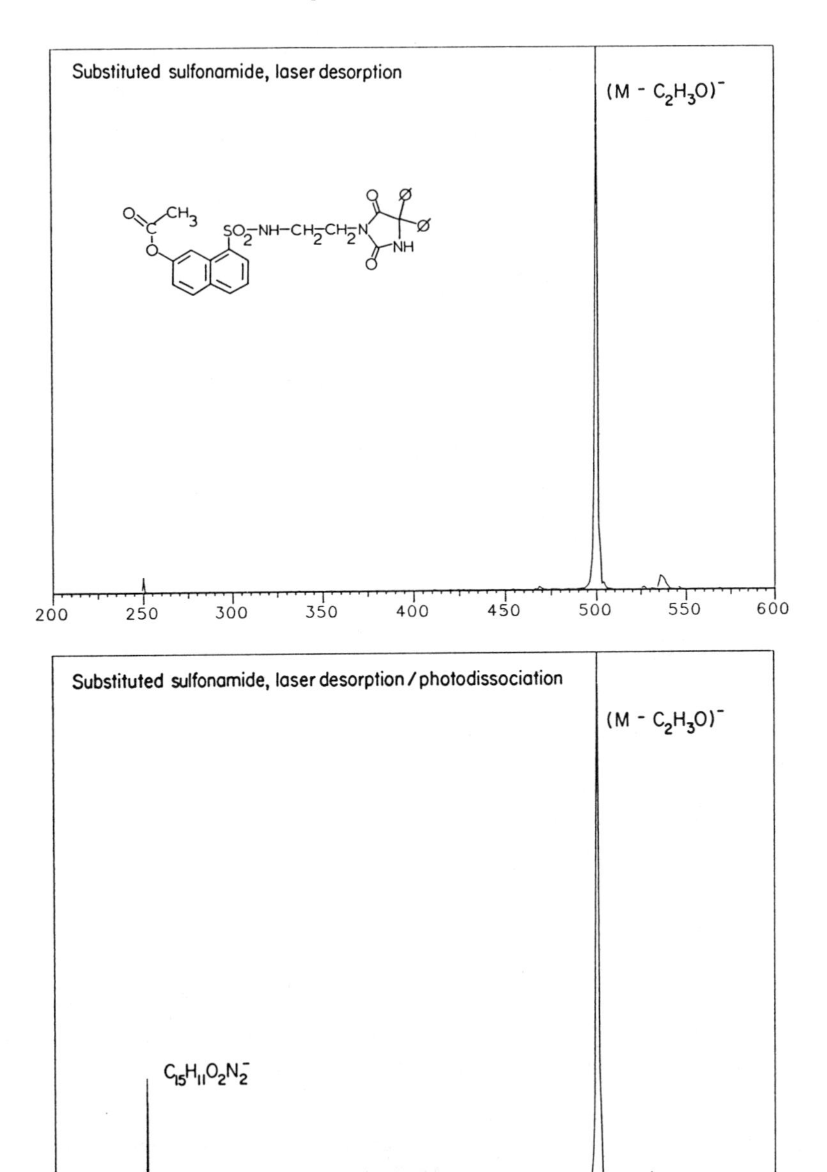


Figure 9. Top. Negative fragment ion formed from the substituted sulfonamide by laser desorption using pulsed $CO_2$ laser. Bottom. Dissociation upon irradiation with a cw $CO_2$ laser, all other conditions identical to those in the top spectrum.
(Reproduced from ref. 18. Copyright 1987 American Chemical Society.)

the use of a Nd:YAG pumped dye laser system for studies similar to those discussed here.

Acknowledgments

We thank Mr. Andy Griffis and Professor Russell Drago for preparation of the CoSALOPH sample, Dr. Robert Weller and Dr. Christopher Riley for donation of the CHIP sample, and Professor Alan Katritzky for the t-butylpyridinium perchlorate sample. This work was supported in part by the Office of Naval Research.

Literature Cited

1. Conzemius, R. J.; Capellen, R. J. Int. J. Mass Spectrom. Ion Phys. 1980, 34, 197.
2. Hillenkamp, F. In Ion Formation from Organic Solids; Benninghoven, A., Ed.; Springer Verlag: Berlin, 1983.
3. Tabet, J.-C.; Cotter, R. J. Int. J. Mass Spectrom. Ion Phys. 1983, 54, 151.
4. Van Breemen, R. B.; Snow, M.; Cotter, R. J. Int. J. Mass Spectrom. Ion Phys. 1983, 49, 35.
5. McCrery, D. A.; Ledford, E. B., Jr.; Gross, M. L. Anal. Chem. 1982, 54, 1435.
6. Wilkins, C. L.; Weil, D. A.; Yang, C. L. C.; Ijames, C. F. Anal. Chem. 1985, 47, 520.
7. Levsen, K.; Schwarz, H. Mass Spectrom. Rev. 1983, 125, 1.
8. Dunbar, R. C. In Gas Phase Ion Chemistry; Bowers, M. T., Ed.; Academic Press: New York, 1984; Vol. 3, p. 130.
9. Thorne, L. R.; Beauchamp, J. L. In Gas Phase Ion Chemistry; Bowers, M. T., Ed.; Academic Press: New York, 1984; Vol. 3, p. 41.
10. Amster, I. J.; Baldwin, M. A.; Cheng, M. T.; Procter, C. J.; McLafferty, F. W. J. Am. Chem. Soc. 1983, 105, 1654.
11. Gilbert, R. G.; Sheil, M. M.; Derrick, P. J. Org. Mass Spectrom. 1985, 20, 430.
12. Bowers, W. D.; Delbert, S.-S.; McIver, R. T. Anal. Chem. 1986, 54, 969.
13. Baykut, G.; Watson, C. H.; Weller, R. R.; Eyler, J. R. J. Am. Chem. Soc. 1985, 107, 8036.
14. Watson, C. H.; Baykut, G.; Battiste, M. A.; Eyler, J. R. Anal. Chim. Acta 1985, 178, 125.
15. Jasinski, J. M.; Rosenfeld, R. N.; Meyer, F. K., Brauman, J. I. J. Am. Chem. Soc. 1982, 104, 652.
16. Van Der Peyl, G. J. Q.; Haverkamp, J.; Kistemaker, Int. J. Mass Spectrom. Ion Phys. 1982, 42, 125.
17. Dunbar, R. C.; Ferrara, J. J. Chem. Phys. 1985, 83, 6229.
18. Watson, C. H.; Baykut, G.; Eyler, J. R. Anal. Chem. 1987, 59, 1133.

RECEIVED June 15, 1987

# Chapter 10

# Gas-Phase Photodissociation of Transition Metal Ion Complexes and Clusters

**Robert L. Hettich [1] and Ben S. Freiser**

**Department of Chemistry, Purdue University, West Lafayette, IN 47907**

Highlights are presented from our recent work on the photodissociation of three categories of transition metal ion species: (I) $ML^+$ (L=ligand), (II) $ML_2^+$, and $MFe^+$ (M=3d metal). The results indicate that these species absorb broadly in the ultraviolet and visible spectral regions. Because of this broad absorption, photodissociation thresholds are attributed to thermodynamic factors and yield absolute bond energies. Isomer differentiation is demonstrated by observing differences in cross sections, spectral band positions, and fragmentation pathways. Interestingly, product ions generated by photodissociation are found in some cases to differ significantly from those produced by collision-induced dissociation. Sophisticated pulse techniques, which permit the multistep synthesis and isolation of a wide variety of organometallic ions, together with the long ion storage times, make FTMS ideally suited for these studies.

Over the past several years, the area of gas-phase transition metal ion chemistry has been gaining increasing attention from the scientific community [1-16]. Its appeal is manifold: first, it has broad implications to a spectrum of other areas such as atmospheric chemistry, corrosion chemistry, solution organometallic chemistry, and surface chemistry; secondly, an arsenal of gas phase techniques are available to study the thermochemistry, kinetics, and mechanisms of these "unusual" species in the absence of such complications as solvent and ligand

[1]Current address: Oak Ridge National Laboratory, P.O. Box X, Oak Ridge, TN 37831

0097–6156/87/0359–0155$06.00/0

effects; and, thirdly, the highly controlled nature of these gas phase experiments permits the study of specific ligand effects and provides data for direct comparison to theory, as new theoretical treatments emerge.

How does the gas-phase chemistry of a bare and, therefore, highly coordinatively unsaturated transition metal ion compare to its solution counterpart? As a representative example, Beauchamp and coworkers proposed Scheme I for the dehydrogenation of butane by $Ni^+$ [2]. The mechanism shown has analogies to solution organo-

Scheme I

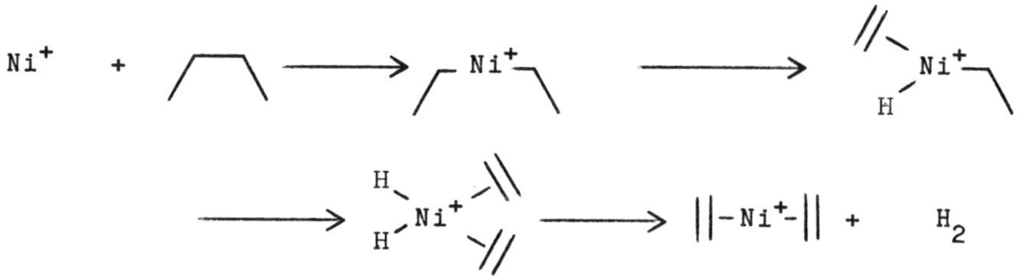

$Ni^+$ + ⟶ $Ni^+$ ⟶ $Ni^+$ H

⟶ H H $Ni^+$ ⟶ $\|-Ni^+-\|$ + $H_2$

metallic chemistry, but is unusual in that oxidative addition of the metal cation is proposed to occur at the C-C bond as opposed to the C-H bond, which is the exact opposite of what is generally observed in solution studies. Oxidative addition of $Fe^+$, $Co^+$, and $Ni^+$ to C-C bonds in longer chain alkanes appears to be general [3-7].

In order to propose a mechanism such as that shown in Scheme I, it is evident that not only must the product ion mass-to-charge ratio (empirical formula) be determined, but also the structure of the ion must be elucidated. Clearly, a different mechanism involving C-H insertion might have been proposed if the $NiC_4H_8^+$ product had been $Ni^+$-butene instead of $Ni(ethene)_2^+$. Several effective structural probes are available to the gas-phase ion chemist including collision-induced dissociation [3,4a,4b], ion-molecule reactions (for metal ion complexes, ligand exchange [1-3] and H/D exchange [5] are commonly employed), and use of specifically labelled neutral compounds [2,6,7]. Each of these methods was applied to the reaction shown in Scheme I and confirm the presence of $Ni(ethene)_2^+$.

Another important consideration in formulating a mechanism is whether each step, as well as the overall process, is thermodynamically feasible. In Scheme I the first step involves cleaving the central C-C bond and forming two $Ni^+$-C bonds. Thus, the determination of metal-ligand bond energies is critically important. Again, several powerful gas-phase techniques can be applied to this problem. Among these, the determination of endothermic reaction thresholds from ion-beam experiments has yielded the majority of the absolute values currently in the literature today [1,8,9]. Ligand displacement reactions and equilibrium measurements have yielded accurate relative metal ion - ligand bond energies for a variety of metal cations and ligands [10,11]. Observation of exothermic ion - molecule reactions can provide limits on these energies and, most recently, competitive ligand loss by collision - induced dissociation has shown promise both for qualitative and quantitative bond information [12].

Despite the successes of the above mentioned techniques for structure and bond energy analyses, inherent uncertainties and limitations in each method make it important to develop new and independent tests for comparison. In this regard our laboratory has initiated an intensive effort to study the gas-phase photodissociation of transition metal containing ions [13-16]. Although a good deal of information is available on the photodissociation of organic ions [17,18], relatively little work has been done on organometallic ions [19,20]. FTMS is ideally suited for these studies because of the wide variety of interesting ions that can be generated and because of the long storage times which permit irradiation of the ions. In this paper some of the highlights of our work is presented together with a description of the methodology and the types of information that can be obtained from these studies.

## Experimental Section

The theory and instrumentation of Fourier transform mass spectrometry (FTMS) have been discussed extensively in this book and elsewhere [21-23]. All experiments were performed on a Nicolet prototype FTMS-1000 Fourier transform mass spectrometer previously described in detail [24] and equipped with a 5.2 cm cubic trapping cell situated between the poles of a Varian 15 in. electromagnet maintained at 0.85 T. The cell was constructed in our laboratory and utilizes two 80% transmittance stainless steel screens as the transmitter plates. This permits irradiation with a 2.5 kW Hg-Xe arc lamp, used in conjunction with a Schoeffel 0.25 m monochromator set for 10 nm resolution. Metal ions are generated by focusing the beam of a Quanta Ray Nd:YAG laser (either the fundamental line at 1064 nm or the frequency doubled line at 532 nm) into the center-drilled hole (1 mm) of a high-purity rod of the appropriate metal supported on the transmitter screen nearest to the laser. The laser ionization technique for generating metal ions has been outlined elsewhere [25].

Details of the collision-induced dissociation (CID) experiments have been described [26]. Argon was used as the collision gas at a total pressure of $\sim 4 \times 10^{-6}$ torr. The collision energy of the ions can be varied (typically between 0 and 100 eV). A Bayard-Alpert ionization gauge was used to monitor static pressures.

Each of the chemicals was obtained commercially and used without further purification, except for multiple freeze-pump-thaw cycles to remove non-condensible gases. Electron impact mass spectrometry indicated no detectable impurities. All samples were admitted to the cell through a General Valve Corp. Series 9 pulsed solenoid valve [27]. The pulsed valve, which was triggered concurrently with the laser pulse used to generate the metal ions, introduced the reagent gas into the vacuum chamber to a maximum pressure of approximately $10^{-5}$ torr. Although the pulse duration of the valve was 2 ms, the high pressure of the reagent gas had a rise time of about 200 ms and was pumped away by a high speed 5 in. diffusion pump in approximately 400 ms. Swept double resonance pulses were then used to isolate the ion of interest, which was subsequently trapped for 3-6 seconds (determined by the ion's cross section for photodissociation) either in the presence or absence of radiation. For each ion, two sets of photodissociation spectra were taken, one at $2 \times 10^{-6}$ torr argon, to permit collisional cooling, and another at a background pressure of $\sim 10^{-8}$ torr [28]. In all cases, data from the collisionally cooled ions are presented.

Photodissociation spectra were obtained by monitoring the appearance of ionic photoproducts as a function of the wavelength of light. Shot-to-shot variation of the laser-generated metal precursor ions made monitoring the photodisappearance of the parent ion impractical. Assuming a one-photon process [17,29], the photodissociation of $AB^+$, eq. 1, can be described by first-order kinetics, eq. 2, where $\sigma_1$ and $\sigma_2$ are the absolute cross sections for

$$AB^+ + h\nu \longrightarrow P_1^+ \quad \text{and} \quad \longrightarrow P_2^+ \tag{1}$$

$$d(AB^+)/dt = -(\sigma_1 + \sigma_2)I(AB^+) \tag{2}$$

production of the photoproducts $P_1^+$ and $P_2^+$, respectively, and I is the photon flux. Integrating eq. 2 and substituting $(AB^+)_o = (AB^+)_{h\nu} + (P_1^+)_{h\nu} + (P_2^+)_{h\nu}$ relates the extent of photodissociation to the cross section at a given wavelength, eq. 3, where $\sigma_t = \sigma_1 + \sigma_2$ is the total cross section and t is the

$$\ln\left[1 + \frac{P_1^+ + P_2^+}{AB^+}\right] = \sigma_t I t \tag{3}$$

irradiation time. Solving eq. 3 for $\sigma_t$ and plotting that value as a function of wavelength, appropriately correcting for blanks (no light), yields the photodissociation spectrum of the ion (i.e., relative $\sigma_t$ vs. wavelength). Clearly, the correct spectrum requires that all of the photodissociation products be detected. Photoappearance curves for the individual photoproducts can be obtained, for example, by solving eq. 4 for $\sigma_1$ and plotting it as a function of wavelength. Each spectrum reported is an

$$\left[\frac{P_1^+}{P_1^+ + P_2^+}\right] \ln\left[1 + \frac{P_1^+ + P_2^+}{AB^+}\right] = \sigma_1 It \qquad (4)$$

average of several trials. The reproducibility of the peak intensities is ± 40% and peak locations is ± 10 nm. Photodissociation thresholds were confirmed by using cut-off filters.

To obtain absolute values for the cross sections of the ions being examined, the photodissociation of $C_7H_8^+$ (from toluene at 20 eV) at 410 nm [$\sigma$(410 nm) = 0.05 $A^2$] [30] was compared to the photodissociation of a given ion at its $\lambda_{max}$, both taken under similar experimental conditions. All cross sections determined in this manner have an estimated uncertainty of ± 50% due to instrumental variables.

## Results and Discussion

### Chemical Systems

As listed below, our initial studies have focussed on three general categories of metal containing ions:

I. $ML^+$ (L = O, S, $CH_2$, $CH_3$, $C_4H_6$, $C_4H_8$, $C_6H_6$, etc.)

II. $ML_2^+$ (L = $C_2H_4$, $C_3H_6$, $C_6H_6$, etc.)

III. $MFe^+$ (M = 3d Series)

Photodissociation of simple metal ion-ligand species (I) can provide information on their absorption characteristics (e.g., is the absorption of light metal or ligand localized or of a charge transfer nature?), structure (particularly in the case of isomers), and bond energy, as discussed in greater depth below. The addition of a second ligand (II) allows one to study the effect of that ligand on the bond energy of the first (i.e., are there any synergistic affects?), as well as its effect on the shape and cross section of the photodissociation spectra. As shown in Scheme I, oxidative addition of a metal ion results in formation of two new bonds to the metal center. In formulating a simple energy diagram for such a process, it is generally assumed that the total ligand bonding energy is the sum of the two individual metal ion-ligand bond energies. While some information is available on multiple ligand systems [10], it is quite rare, and photodissociation holds promise for greatly expanding the data base on this question. Finally, the area of clusters is a rapidly growing one. Our effort in this area has been to develop a method for synthesizing *in situ* a wide variety of dimer (III) [31-33] and trimer ions [34] of known composition and study their gas-phase chemistry.

Once again, photodissociation studies can provide complementary information on these chemically important species.

Specific Ion Synthesis

The unique feature of FTMS to control both the ion and neutral populations by using programmable pulse sequences makes it an extremely powerful tool for generating in a stepwise fashion specific organometallic fragment ions for subsequent chemical and photochemical studies. Figure 1 shows a typical pulse sequence which begins with an ionization pulse. Concurrently with the laser pulse, a pulsed valve fires to admit a burst of reagent gas. The metal ions react with the reagent gas to form the ion of interest, either directly in a primary or secondary reaction, or by a subsequent step such as collision-induced dissociation. For example, $CoOH^+$ can be made by a direct primary reaction of $Co^+$ with $CH_3ONO$ (reaction 5) [13], while metal dimers of the form $MFe^+$ can be synthesized

$$Co^+ + CH_3ONO \longrightarrow CoOH^+ + CH_2NO \quad (5)$$

in a stepwise fashion from $Fe(CO)_5$ by first generating $MFe(CO)_n^+$ followed by collision-induced dissociation to generate the bare $MFe^+$, reactions 6 and 7 [31-33]. Next, a series of ion ejection pulses are used to isolate the

$$M^+ + Fe(CO)_5 \longrightarrow MFe(CO)_{5-X}^+ + XCO \quad (6)$$

$$MFe(CO)_{5-X}^+ \xrightarrow{CID} MFe^+ + (5-X)CO \quad (7)$$

ion of interest. By this time the reagent gas has been pumped away, permitting the ions to be stored efficiently for the relatively long trapping times used for photodissociation. Following an appropiate delay period (with or without light), detection yields the full mass spectrum, and finally a quench pulse eliminates all of the ions from the cell and the whole sequence is repeated.

Information from Photodissociation

In order to observe photodissociation process 1, three criteria must be met: first, the ion must absorb a photon; second, the photon must have sufficient energy to cause fragmentation; and third, the quantum yield for photodissociation must be non-zero. Figure 2 is useful in understanding the information inherent in a photodissociation experiment. If the first allowed electronic state of $AB^+$ lies at an energy above that required to generate the products $P_1^+$ and $P_2^+$ (left side of Figure 2), the observed photodissociation onset is spectroscopically determined and yields only an upper energy limit for the processes in reaction 1. In other words, even if there is sufficient energy in the photon to cause ion fragmentation, clearly the process will not occur if the ion does not absorb the photon. Alter-

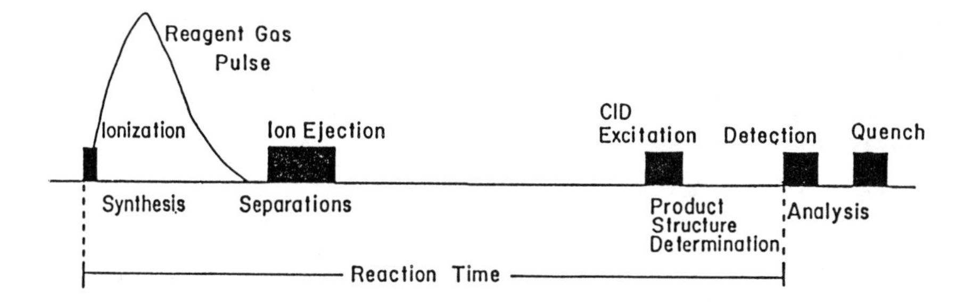


Figure 1. Pulse sequence used to synthesize, isolate, and study specific metal-containing ions.

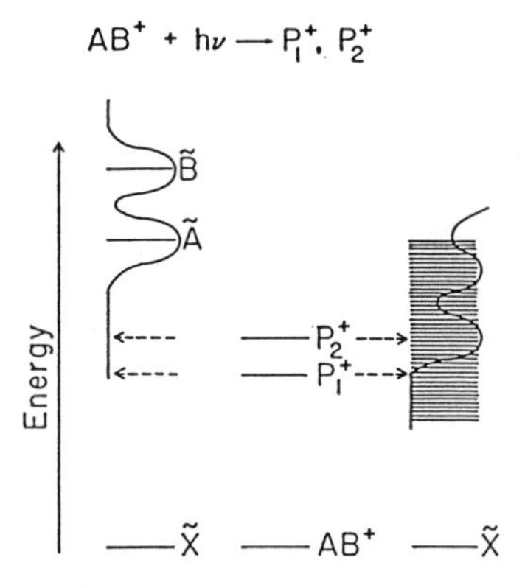


Figure 2. Energy level diagram depicting cases where photodissociation thresholds are determined by spectroscopic factors (left side) and thermodynamic factors (right side).

natively, if an allowed electronic state fortuitously lies at the same energy as the lowest energy process or, as shown in Figure 2 (right side), there is a high density of allowed states having energies in the vicinity of the thermodynamic threshold, the observed photodissociation onset should accurately reflect the dissociation energy for process 1 provided the quantum yield is non-zero. Onsets for higher energy products will also reflect the thermochemistry if rapid internal conversion to a vibrationally excited ground state occurs randomizing the energy. Thus, if an ion absorbs light over a broad range, say from the ir to the uv, then photodissociation can begin to occur when the photon energy is sufficient to cause fragmentation. In any event, the photodissociation spectrum is an indirect measure of the true gas-phase absorption spectrum and, as such, provides a "fingerprint" of the ground state structure of the ion.

In general, the state diagram on the left in Figure 2 is a good description for organic ions. As an example, the photodissociation spectrum of benzoyl cation obtained by monitoring process 8 has two maxima at 260 nm and 310 nm

$$C_6H_5CO^+ + h\nu \longrightarrow C_6H_5^+ + CO \quad (8)$$

attributed to the transitions $^1A_{1g}$ $^1B_{1u}$ and $^1A_{1g}$ $^1B_{2u}$, respectively, and an onset for photodissociation at 350 nm or 3.5 eV [18]. The onset is considerably higher than the actual enthalpy of 2.3 eV required to decarbonylate the benzoyl ion.

In contrast to the organic ions, all of our work to date on the chemical systems I, II, and III, suggest that these metal complexes have a high density of low lying electronic states and, therefore, absorb broadly yielding photodissociation onsets which reflect the thermochemistry. In particular, values obtained by observing photodissociation onsets are found in general to be in good agreement with those obtained by other techniques [15]. Certainly, exceptions to this will be found. Finally, although the information content of these photodissociation spectra is great with regard to ion structure and thermochemistry, the high density of low lying electronic states makes band assignments virtually impossible at this time. In fact such an interpretation will provide a real challenge to theorists for years to come.

(I) Photodissociation of $ML^+$

One of the first examples from our laboratory [13] which suggested that photodissociation thresholds could yield quantitative metal-ligand bond energies was from a comparison of the photodissociation spectra of $FeOH^+$ and $FeCO^+$ obtained by monitoring reactions 9 and 10, respectively, and shown in Figure 3. The two spectra are remarkably similar with two absorption maxima observed

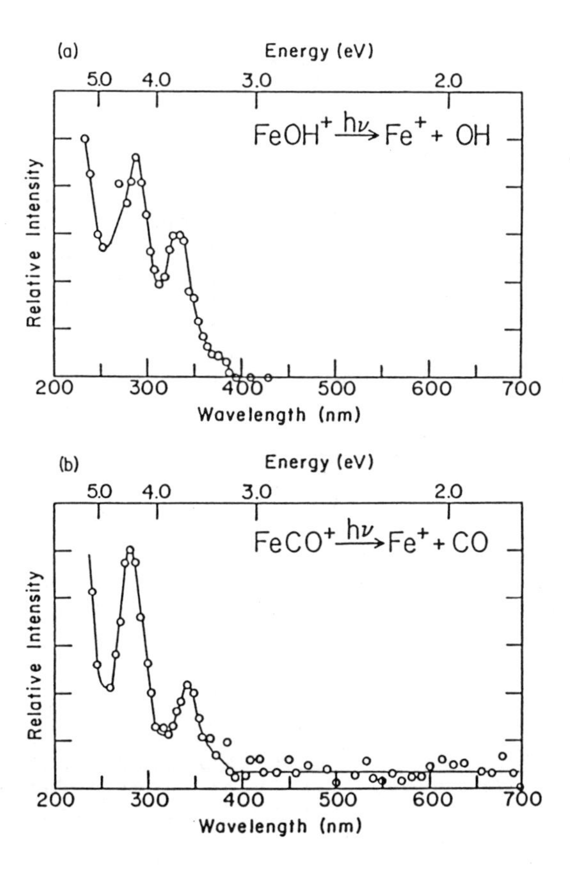


Figure 3. (a) Photodissociation spectrum of $FeOH^+$ obtained by monitoring reaction 9 as a function of wavelength. No photodissociation is observed at wavelengths greater than 390 nm. (b) Photodissociation spectrum of $FeCO^+$ generated by electron impact on $Fe(CO)_5$. The observation of photodisssociation at wavelengths greater than 660 nm was confirmed by using a cutoff filter.
(Reproduced from ref. 13. Copyright 1984 American Chemical Society.)

$$FeOH^+ + h\nu \longrightarrow Fe^+ + OH \qquad (9)$$

$$FeCO^+ + h\nu \longrightarrow Fe^+ + CO \qquad (10)$$

near 290 nm and 335 nm, suggesting that the transitions are metal localized. The $FeOH^+$ spectrum, however, has a photodissociation threshold of 390 ±10 nm, while $FeCO^+$ is observed to have a low intensity long wavelength absorption tail from about 400 to 700 nm. The cutoff observed at 390 nm for $FeOH^+$ yields an absolute bond energy $D^o(Fe^+-OH)$ = 73±3 kcal/mol, which is in excellent agreement with a value of 76±5 kcal/mol obtained by measuring the ionization potential of FeOH [35] and a value of 77±8 kcal/mol obtained by bracketing the proton affinity of FeO [13]. Appearance potential measurements on $FeCO^+$ from $Fe(CO)_5$ were interpreted to yield $D^o(Fe^+-CO)$ = 60 kcal/mol [36]. These same results, however, were later reinterpreted to yield $D^o(Fe^+-CO)$ = 38 kcal/mol [37]. Observation of photodissociation reaction 10 at ~700 nm or 41 kcal/mol provides additional support for the lower value.

As shown in Table I, the evidence is mounting that photodissociation thresholds do in fact yield accurate metal-ligand bond energies. If there is a disagreement between the photodissociation value and the value from an alternative technique, the photodissociation value tends to be lower in contrast, for example, to the benzoyl cation case discussed above. Arriving at a lower value by photodissociation suggests the possibility that the precursor ion may be generated with excess internal energy [4b,38]. To circumvent this problem, once formed the precursor ions are permitted to undergo thermalizing collisions with argon (see experimental section) and the effect on the threshold, if any, is noted. In addition, whenever possible, the precursor ion of interest was generated from more than one neutral source. For the majority of the systems studied, however, neither the collisional cooling step nor the synthesis from different neutral precursors had a measureable effect on the threshold. One notable acception is $FeCH_2^+$ as discussed below.

Transition-metal methylidenes have been implicated as intermediates in a variety of important catalytic transformations including olefin metathesis, the Ziegler-Natta polymerization of olefins, olefin homologation, and the heterogeneous Fischer-Tropsch process. Likewise, the increasing literature on bare $MCH_2^+$ in the gas phase has shown these species to have interesting physical and chemical properties [39-42]. It is perhaps not surprising, therefore, that the photochemistry of these species has proven to be among the most interesting.

Our initial work on $MCH_2^+$ was for M = Fe and Co [14], but these studies have recently been expanded to include M = Rh, Nb, and La [16]. The methylidenes can be

Table I. Bond Energy Determinations

| $A^+$-B | $D^o(A^+-B)$(kcal/mol) photodissociation | other |
|---|---|---|
| $Fe^+-CH_2$ | 82±5(a) | 96±5 (b) |
| $Fe^+-CH$ | 101±5(a) | 115±20 (c) |
| $Fe^+-C$ | 94±5(a) | 89 (c) |
| $Co^+-CH_2$ | 84±5(a) | 85±7 (d) |
| $Co^+-CH$ | 100±5(a) | |
| $Co^+-C$ | 90±5(a) | 98 (c) |
| $Nb^+-CH_2$ | 109±7(d) | 112 (d) |
| $Nb^+-CH$ | 145±8(d) | |
| $Nb^+-C$ | >138 (d) | |
| $Rh^+-CH_2$ | 91±5(d) | 94±5 (e) |
| $Rh^+-CH$ | 102±7(d) | |
| $Rh^+-C$ | >120 (d) | 164±16 (d) |
| $La^+-CH_2$ | 106±5(d) | 106±5 (d) |
| $La^+-CH$ | 125±8(d) | |
| $La^+-C$ | 102±8(d) | |
| $Fe^+-CH_3$ | 65±5(f) | 69±5 (g) |
| $Co^+-CH_3$ | 57±7(f) | 61±4 (h) |
| $Fe^+-O$ | 68±5(f) | 68±3 (h) |
| $Fe^+-S$ | 65±5(f) | 74>x>59(i) |
| $Co^+-S$ | 62±5(f) | 74>x>59(i) |
| $Ni^+-S$ | 60±5(f) | |
| $V^+-C_6H_6$ | 62±5(f) | |
| $C_6H_6V^+-C_6H_6$ | 57±5(f) | |
| $Fe^+-C_6H_6$ | 55±5(f) | 59±5 (j) |
| $Co^+-C_6H_6$ | 68±5(f) | 71>x>61(k) |
| $Sc^+-Fe$ | 48±5(l) | 49±6 (m) |
| $Ti^+-Fe$ | 60±6(l) | >49 (l) |
| $V^+-Fe$ | 75±5(n) | >62 (f) |
| $Cr^+-Fe$ | 50±7(l) | |
| $Fe^+-Fe$ | 62±5(l) | 58±7 (o) |
| $Co^+-Fe$ | 62±5(l) | 62±6 (o) |
| $Ni^+-Fe$ | 64±5(l) | <68±5 (p) |
| $Cu^+-Fe$ | 53±7(l) | >52 (l) |
| $Nb^+-Fe$ | 68±5(l) | >60 (q) |
| $Ta^+-Fe$ | 72±5(l) | >60 (q) |

(a) Reference 14. (b) Reference 40. (c) Beauchamp, J.L., private communication. (d) Armentrout, P.B.; Beauchamp, J.L. J. Chem. Phys. 1981, 74, 2819. (e) Jacobson, D.B. Freiser, B.S. J. Am. Chem. Soc. 1985, 107, 5870. (f) Reference 15. (g) Reference 37. (h) Reference 8. (i) Carlin, T.J. Ph.D. Thesis, Purdue University, 1984. (j) Jacobson, D.B.; Freiser, B.S. J. Am. Chem. Soc. 1984, 106, 3900. (k) Reference 31. (l) Reference 16. (m) Lech, L.M.; Freiser, B.S. unpublished results. (n) Reference 33. (o) Reference 31. (p) Jacobson, D.B.; Freiser, B.S. unpublished results. (q) Buckner, S.; MacMahon, T.; Freiser, B.S. unpublished results.

formed from a variety of different neutral precursors depending on the overall thermochemistry. In fact the reaction or lack of reaction of the metal ions with the compounds used, which included cycloheptatriene (70 kcal/mol), ethylene oxide (78 kcal/mol), cyclopropane (93 kcal/mol), propene (101 kcal/mol), and methane (112 kcal/mol) where the energy shown in parenthesis is that required to remove a $CH_2$ unit [43], provided the first indication of their metal ion - methylidene bond energies. Photodissociation of these $MCH_2^+$ species proceeds by three pathways, reactions 11-13. These results were un-

$$MCH_2^+ + h\nu \rightarrow \begin{cases} M^+ + CH_2 & (11) \\ MC^+ + H_2 & (12) \\ MCH^+ + H & (13) \end{cases}$$

expected in the first studies on $FeCH_2^+$ and $CoCH_2^+$, because they were in direct contrast to the low-energy (0-100 ev) collision-induced dissociation results yielding exclusive cleavage of $CH_2$ to form $M^+$, and because both $FeCH_3^+$ and $CoCH_3^+$ photodissociate by $CH_3$ loss, exclusively. A similar difference between the photodissociation and collision-induced dissociation pathways was observed for $LaCH_2^+$, whereas observation of both the carbide as well as the bare metal ion following collision-induced dissociation of $RhCH_2^+$ and $NbCH_2^+$ more closely parallels the photodissociation pathways observed for these ions. By monitoring the photodissociation thresholds for each of the three photoproducts in reactions 11-13, bond energies for $M^+$- C, $M^+$-CH, and $M^+$-$CH_2$ were obtained (see Table I).

One of the striking disagreements between a bond energy determined by photodissociation and one from the ion-beam experiment is $D^o(Fe^+-CH_2)$ where photodissociation yields 82±5 kcal/mol [14] compared to the 96±5 kcal/mol reported earlier [40]. As mentioned in one of the above sections, a possible negative determinate error in the photodissociation experiment can arise if the precursor ion is internally excited prior to photon absorption. If so, collisional relaxation of the ions should result in a higher energy onset. Figure 4 shows the long wavelength threshold region for $FeCH_2^+$ photodissociating to $Fe^+$, taken at different pressures, which is the "worst case" we have encountered to date. Observation of photodissociation beyond 400 nm in Figure 4 would imply $D^o(Fe^+-CH_2) < 71$ kcal/mol which is unreasonably low, since the reaction of $Fe^+$ with ethylene oxide to form $FeCH_2^+$ requires a bond energy of 78 kcal/mol or greater [43]. Adding argon and raising the pressure clearly results in a decrease of dissociation in the tail region (> 350 nm) and a slight enhancing of the peak maximum near 330 nm. While the significant pressure quenching may indicate the presence of internally or kinetically excited $FeCH_2^+$ ions, an alternative explanation is that the ion is absorbing two photons (multiphoton process) [29] although this is believed to be less likely. Presumably, a sufficiently high pressure of Ar would, in any event, completely quench the tail, but reduced trapping efficiency precluded doing this experiment. The wavelength at which Ar begins to reduce the photoappearance of $Fe^+$ and which corresponds to the extrapolated onset of the absorption band is about 350 nm or 82 kcal/mol. The earlier reported value of $D^o(Fe^+-CH_2) = 96\pm5$ kcal/mol seems high on the basis of the intense photoappearance of $Fe^+$ at 340 nm even in the presence of Ar, which requires $D^o(Fe^+-CH_2) < 84$ kcal/mol. The photodissociation result, however, is consistent with the 77-87 kcal/mol range suggested from ion-molecule reaction studies [41a]. This example demonstrates the value of applying a variety of techniques to determine these important thermodynamic parameters. Finally, we are continuing to study the underlying reasons why methylidenes are particularly susceptible to these long wavelength effects.

Two isomers of $CoC_5H_{10}^+$ which can be postulated as the products of various ion-molecule reactions [3] are shown by structures I and II. Structure I can easily rearrange to structure II upon activation, as is evident

Co⁺ -(1-pentene) (I)    (ethylene)- Co⁺ -(propene) (II)

from the primary reactions of $Co^+$ with 1-pentene, specifically reactions 16 and 17. These same products are observed in the CID of $Co^+$-pentene, with the exception

$$Co^+ + \text{1-pentene} \longrightarrow \begin{cases} \xrightarrow{7\%} CoC_5H_8^+ + H_2 & (14) \\ \xrightarrow{10\%} CoC_4H_6^+ + CH_4 & (15) \\ \xrightarrow{62\%} CoC_3H_6^+ + C_2H_4 & (16) \\ \xrightarrow{21\%} CoC_2H_4^+ + C_3H_6 & (17) \end{cases}$$

that only a small amount of $CoC_5H_8^+$ was observed [3].

The ion of structure I can presumably be made by reaction 18 [3]. Displacement of propene should allow the ionic product of reaction 18 to be generated with

$$Co^+\text{-propene} + \text{1-pentene} \longrightarrow Co^+\text{-(1-pentene)} + \text{propene} \qquad (18)$$

little excess internal energy, preferentially favoring structure I over structure II.

Photodissociation of $CoC_5H_{10}^+$ from reaction 18, structure IV, generates five photoproducts, reactions 19-23. The percentages of products were taken at $\lambda_{max}$ =

$$Co^+\text{-pentene} + h\nu \longrightarrow \begin{cases} \xrightarrow{2\%} CoC_5H_8^+ + H_2 & (19) \\ \xrightarrow{3\%} CoC_4H_6^+ + CH_4 & (20) \\ \xrightarrow{62\%} CoC_3H_6^+ + C_2H_4 & (21) \\ \xrightarrow{11\%} CoC_2H_4^+ + C_3H_6 & (22) \\ \xrightarrow{22\%} Co^+ + C_5H_{10} & (23) \end{cases}$$

320 nm. The photodissociation spectrum of $Co^+$-pentene, Figure 5, indicates peak maxima at 320 nm ($\sigma$ = 0.02 $A^2$) and 370 nm. All five photoproducts are observed at wavelengths out to at least 430 nm. If the photo-appearance of $Co^+$ is due solely to reaction 23, then observation of $Co^+$ at 430 nm would imply $D^o(Co^+-C_5H_{10})$ < 66 kcal/mol. This value must be expressed with caution, however, since the primary photoproducts can also further dissociate to give $Co^+$.

## II. Photodissociation of $ML_2^+$

The ion of structure II can be made presumably by reaction 24 [3]. The high static pressure of argon

$$Co^+ + \text{butene} \xrightarrow{-H_2} Co^+\text{-propene} \xrightarrow[C_2H_4]{Ar} (\text{ethylene})\text{-}Co^+\text{-}(\text{propene}) \qquad (24)$$

stabilizes the condensation of $C_2H_4$ onto $CoC_3H_6^+$. Photodissociation of $CoC_5H_{10}^+$ from reaction 24 (Figure 6) yields only three photoproducts, reactions 25-27, which are all observed at least out to 430 nm. The relative abundances of these photoproducts were taken at

$$(\text{ethylene})\text{-}Co^+\text{-}(\text{propene}) + h\nu \longrightarrow \begin{cases} \xrightarrow{61\%} CoC_3H_6^+ + C_2H_4 & (25) \\ \xrightarrow{15\%} CoC_2H_4^+ + C_3H_6 & (26) \\ \xrightarrow{24\%} Co^+ + (C_5H_{10}) & (27) \end{cases}$$

$\lambda_{max}$ = 320 nm($\sigma$= 0.05 $A^2$). No $CoC_4H_6^+$ or $CoC_5H_8^+$ are observed, supporting the fact that this ion exists as structure II. The spectra of these two isomers I and II are are similar, but the 370 nm peak observed in the spectrum of $Co^+$- pentene is absent in the spectrum of $Co^+$-(propene) (ethylene). The enhanced photodissociation

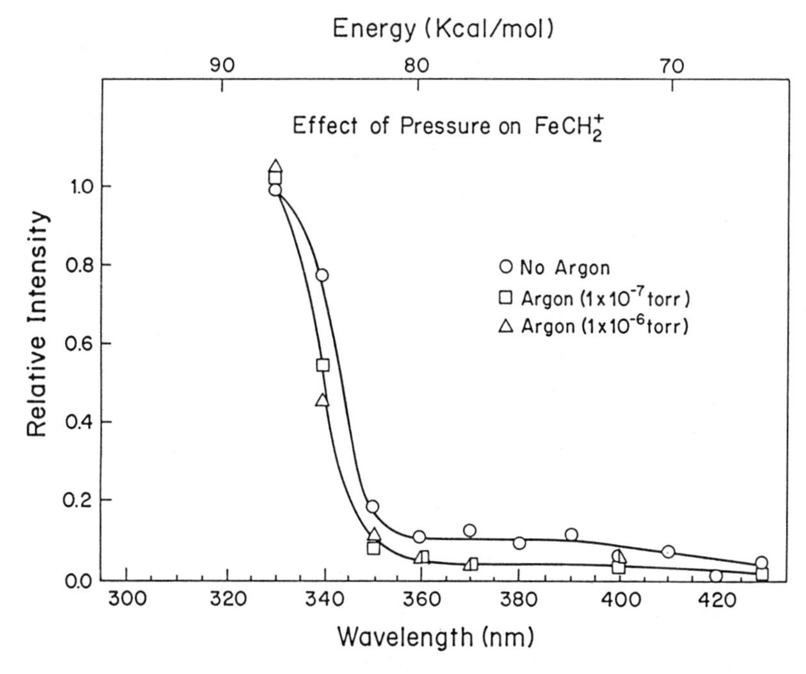


Figure 4. Long wavelength threshold region for $FeCH_2^+$ photodissociating to $Fe^+$ at various pressures.

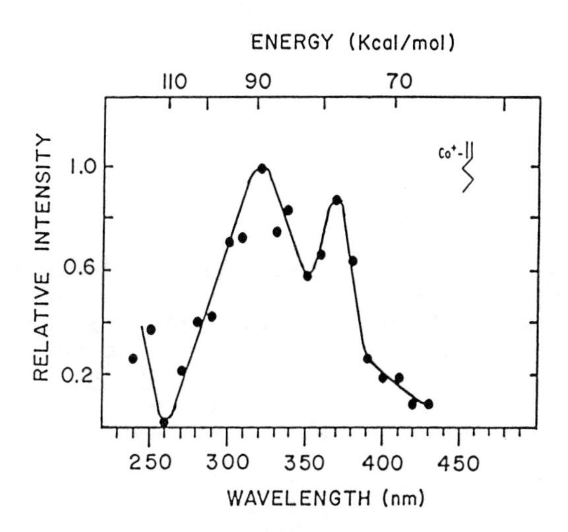


Figure 5. Photodissociation spectrum of $CoC_5H_{10}^+$ generated from reaction 18.

cross section observed for structure II (ie. 0.05 $Å^2$) is typical of a trend we have observed that photodissociation cross section tends to increase with the addition of a second ligand [15]. The difference in cross sections alone, however, would not be sufficient to distinguish the two isomers. A photoappearance threshold for $Co^+$, reaction 27, could not be obtained due to the secondary dissociation of $CoC_3H_6^+$ and $CoC_2H_4^+$ to yield $Co^+$.

Recently we reported that $Ni(C_2H_4)_2^+$ could also be distinguished from three other isomers ($Ni^+$-butene, $Ni^+$-isobutene, nickelacyclopentane cation) on the basis of its unique photodissociation spectrum and photoproducts [15]. In particular $Ni(C_2H_4)_2^+$ photodissociates to give two products of about equal abundance (by loss of $C_2H_4$ and $C_4H_8$). Continuous ejection of $Ni^+-C_2H_4$ allows reaction 28 to be monitored. As shown in Figure 7,

$$|| - Ni^+ - || \; + \; h\nu \longrightarrow Ni^+ \; + \; 2\ C_2H_4 \qquad (28)$$

reaction 28 is observed for < 370 nm implying $D^o(Ni^+-2C_2H_4)$ = 77±5 kcal/mol. The fact that this value is not significantly different than twice $D^o(Ni^+-C_2H_4)$ = 37±2 kcal/mol suggests that synergistic effects are not very pronounced in this ion. While this may prove to be typical, it is not expected to be the rule. For example, preliminary studies on the bis-benzene ions, $M(C_6H_6)_2^+$, suggest a negligible synergistic effect for M = V [15] and Sc [44], a negative effect for $La^+$(i.e., $D^o(La^+-C_6H_6) > D^o(LaC_6H_6^+-C_6H_6)$, and a positive effect for $Cu^+$ (i.e., $D^o(Cu^+-C_6H_6) < D^o(CuC_6H_6^+ - C_6H_6)$ [44].

## III. Photodissociation of $MFe^+$

Examination of the bonding and energetics of small bare clusters has become a topic of considerable interest in recent years. An exciting array of methods has been developed to generate clusters of varying sizes and composition for study in the gas phase. Some of these new methods include sputtering techniques [45], gas evaporation techniques [46], supersonic expansion techniques with oven [47] and pulsed laser sources [48], and multiphoton dissociation of multinuclear organometallic compounds [49]. As described above, our laboratory demonstrated the synthesis of small homonuclear and heteronuclear transition metal cluster ions *in situ* by using FTMS [31-33]. For example, reactions 6 and 7 have been used to synthesize a wide variety of $MFe^+$ species. The clear advantages of this method are that not only is there a great deal of selectivity in generating specific clusters, but once formed the full power of FTMS can be applied to study the chemistry and photochemistry of these species in detail. This approach is typified by extensive studies on the reactivities of $CoFe^+$ [32] and $VFe^+$ [33] with alkenes. In a soon to be published

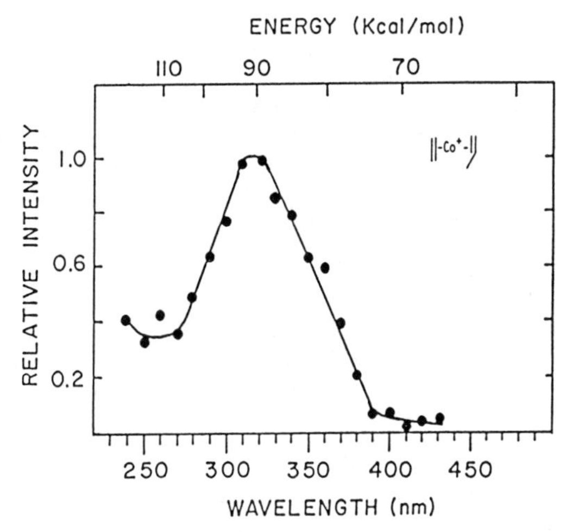


Figure 6. Photodissociation spectrum of $CoC_5H_{10}^+$ generated from reaction 24.

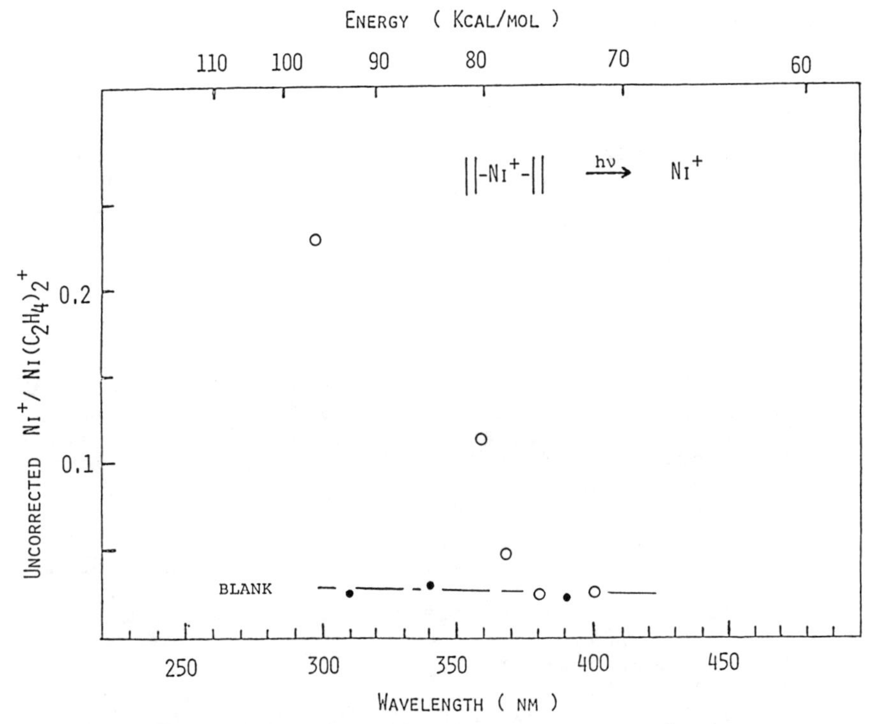


Figure 7. Threshold region for $Ni(C_2H_4)_2^+$ photodissociating to $Ni^+$. In this experiment, $NiC_2H_4^+$ was continuously ejected.

paper [16], we expand on this work by reporting on the photodissociation of $MFe^+$ (M = Sc, Ti, V, Cr, Fe, Co, Ni, Cu, Nb, Ta). To summarize those results, both $M^+$ and $Fe^+$ are observed as photoproducts, with the metal having the lowest ionization potential predominating. The photodissociation spectra reveal broad absorption in the ultraviolet and visible regions with a range of cross sections from 0.06 $Å^2$ for $VFe^+$ to 0.62 $Å^2$ for for $CrFe^+$. Bond energies obtained by observing photoappearance onsets are in the range of 48 kcal/mol for $ScFe^+$ to 75 kcal/mol for $VFe^+$ (see Table I). These studies can be readily extended to other dimer series, such as $MV^+$, $MCr^+$, and $MCo^+$ generated in analogy to reactions 6 and 7 from $V(CO)_6$, $Cr(CO)_6$, and $Co(CO)_3NO$, respectively, as well as to trimers [34] and higher order clusters. Clearly, the surface has just been scratched!

Acknowledgment is made to the Division of Chemical Sciences in the Office of Basic Energy Sciences in the United States Department of Energy (DE-AC02-80ER10689) for supporting this research and to the National Science Foundation (CHE-8310039) for continued support of FTMS methodology. The authors also wish to thank Michelle Buchanan for her invitation to contribute to this symposium.

## References

1. For a comprehensive review on gas-phase metal ion chemistry see: Allison, J. in Progress in Inorganic Chemistry, Ed. Lippard, S. J., Wiley - Interscience, New York, Vol. 34, 628, 1986.
2. Halle, L.F.; Houriet, R.; Kappes, M.; Staley, R.H.; Beauchamp, J.L. J. Am. Chem. Soc. 1982, 104, 6293.
3. Jacobson, D.B.; Freiser, B.S.; J. Am. Chem. Soc. 1983, 105, 5197.
4. (a) Larsen, B.S.; Ridge, D.P. J. Am. Chem. Soc. 1984, 106, 1912. (b) Freas, R.B.; Ridge, D.P. J. Am. Chem. Soc., 1980, 102, 7129. (c) Armentrout, P.B.; Beauchamp, J.L.; J. Am. Chem. Soc., 1980, 102, 1736.
5. Jacobson, D.B.; Freiser, B.S. J. Am. Chem. Soc. 1985, 107, 72.
6. Houriet, R.; Halle, L.F.; Beauchamp, J.L. Organometallics 1983, 2, 1818.
7. Allison, J.; Freas, R.B.; Ridge, D.P., J. Am. Chem. Soc., 1979, 101, 1332.
8. Armentrout, P.B.; Beauchamp, J.L. J. Am. Chem. Soc. 1981, 103, 784.
9. Aristov, N.; Armentrout, P.B. J. Am. Chem. Soc. 1984, 106, 4065.
10. Jones, R.W.; Staley, R.H. J. Am. Chem. Soc. 1982, 104, 2296.
11. Uppal, J.S.; Staley, R.H. J. Am. Chem. Soc. 1982, 104, 1235.

12. McLuckey, S.A., Schoen, A.E.; Cooks, R.G. J. Am. Chem. Soc. 1982, 104, 848.
13. Cassady, C.J.; Freiser, B.S. J. Am. Chem. Soc. 1984, 106, 6176.
14. Hettich, R.L.; Freiser, B.S. J. Am. Chem. Soc. 1986, 108, 2537.
15. Hettich, R.L.; Jackson, T.C.; Stanko, E.M.; Freiser, B.S. J. Am. Chem. Soc. 1986, 108, 5086.
16. Hettich, R.L.; Freiser, B.S. J. Am. Chem. Soc., in press.
17. Dunbar, R.C. "Gas Phase Ion Chemistry;" Bowers, M.T., Ed.; Academic Press, Inc.; New York, 1984; Vol. 3, Chap. 20.
18. Freiser, B.S.; Beauchamp, J. L. J. Am. Chem. Soc. 1976, 98, 3136.
19. Dunbar, R.C.; Hutchinson, B.B. J. Am. Chem. Soc. 1974, 96, 3816.
20. Burnier, R.C.; Freiser, B.S. Inorg. Chem. 1979, 18, 906.
21. Comisarow, M.B. Adv. Mass Spec. 1980, 8, 1698.
22. Gross, M.L.; Rempel, D.L. Science 1984, 226, 261.
23. Lande, Jr.,D.A.; Johlman, C.L.; Brown, R.S.; Weil, D.A.; Wilkins, C.L. Mass Spec. Rev. 1986, 5, 107.
24. Cody, R.B.; Burnier, R.C.; Freiser, B.S. Anal. Chem. 1982, 54, 96.
25. Cody, R.B.; Burnier, R.C.; Reents, Jr., W.D.; Carlin, T.J.; McCrery, D.A.; Lengal, R.K.; Freiser, B.S. Int. J. Mass Spec. Ion Phys. 1980, 33, 37.
26. Cody, R.B.; Burnier, R.C.; Freiser, B.S. Anal. Chem. 1982, 54, 96.
27. Carlin, T.J.; Freiser, B.S. Anal. Chem. 1983, 55, 571.
28. The photodissociation spectra of all of the ions showed no pressure dependence, except for slight quenching of the low energy tail region, indicating that photodissociation in these cases is probably due to one-photon excitation and that the precursor ions do not have substantial internal energy.
29. Freiser, B.S.; Beauchamp, J.L. Chem. Phys. Lett. 1975, 35, 35.
30. Dunbar, R.C. Chem. Phys. Lett. 1975, 32, 508.
31. Jacobson, D.B.; Freiser, B.S. J. Am. Chem. Soc. 1984, 106, 4623.
32. Jacobson, D.B.; Freiser, B.S. J. Am. Chem. Soc. 1985, 107, 1581.
33. Hettich, R.L.; Freiser, B.S. J. Am. Chem. Soc. 1985, 107, 6222.
34. Jacobson, D.B.; Freiser, B.S. J. Am. Chem. Soc. 1984, 106, 5351.
35. Murad, E. J. Chem. Phys. 1980, 73, 1381.
36. Distefano, G.J. Res. Natl. Bur. Stand. Sect. A. 1970, 74A, 233.
37. Halle, L.F.; Armentrout, P.B.; Beauchamp, J.L. Organometallics 1982, 1, 963.

38. Metal ions can be made with excess internal and/or kinetic energy. See, for example, (a) Kang, H.; Beauchamp, J.L. J. Phys. Chem. 1985, 89, 3364. (b) Halle, L.F.; Armentrout, P.B.; Beauchamp, J.L. J. Am Chem. Soc., 103, 962 (1981).
39. (a) Stevens, A.E.; Beauchamp, J.L. J. Am. Chem. Soc. 1980, 100, 2584. (b) Stevens, A.E.; Beauchamp, J.L. J. Am. Chem. Soc. 1979, 101, 6449.
40. Armentrout, P.B.; Halle, L.F.; Beauchamp. J.L. J. Am. Chem. Soc. 1981, 103, 6501.
41. (a) Jacobson, D.B.; Freiser, B.S. J. Am. Chem. Soc. 1985, 107, 2605. (b) Jacobson, D.B.; Freiser, B.S. J. Am. Chem. Soc. 1985, 107, 4375.
42. (a) Carter, A.E.; Goddard III, W.A. J. Am. Chem. Soc. 1986, 108, 2180. (b) Shim, I.; Gingerich, K.A. J. Chem. Phys. 1982, 76, 3833.
43. All heats of formation (and other supplementary values) are taaken from: Rosenstock, H.M.; Draxl, K.; Steiner, B.W.; Herron, J.T. J. Phys. Chem. Ref. Data. Suppl. 1 1977, 6.
44. Lech, L.M.; Tews, E.C.; Huang, Y.; Freiser, B.S. unpublished results.
45. (a) Katakuse, I.; Ichihara, T.; Fujjita, Y.; Matsuo, T.; Sakurai, T.; Matsuda, H. Int. J. Mas Spec. Ion Proc. 1985, 67, 229. (b) Freas, R.B.; Campana, J.E. J. Am. Chem. Soc. 1985, 107, 6202.
46. (a) Sattler, K.; Muhlbach, J.; Recknagel, E. Phys. Rev. Lett. 1980, 45, 821. (b) Abe, H.; Schulze, W.; Tesche, B. Chem. Phys. 1980, 47, 95. (c) Godenfeld, I.; Frank, F.; Schulze, W.; Winter, B. Int. J. Mass Spec. Ion Proc. 1986, 71, 103.
47. (a) Riley, S.J.; Parks, E.K.; Mao, C.R.; Poppo, L.G.; Wexler, S. J. Phys. Chem. 1982, 86, 3911. (b) Bowles, R.S.; Park, S.B.; Otsuka, N.; Andres, R.P. J. Mol. Catal. 1983, 20, 279.
48. (a) Bondybey, V.E.; English, J.H. J. Chem. Phys. 1982, 74, 6978. (b) Morse, M.D.; Hansen, G.P.; Langridge-Smith, P.R.R.; Zheng, L.S.; Geusic, M.E.; Michalopoulos, D.L.; Smalley, R.E. J. Chem. Phys. 1984, 80, 5400. (c) Smalley, R.E. Laser Chem. 1983, 2, 167.
49. Leopold, D.G.; Vaida, V. J. Am. Chem. Soc. 1983, 105, 6809.

RECEIVED May 12, 1987

# Chapter 11

# Fourier Transform Mass Spectrometry Studies of Negative Ion Processes

**Michelle V. Buchanan and Marcus B. Wise**

**Analytical Chemistry Division, Oak Ridge National Laboratory, Oak Ridge, TN 37831-6120**

FTMS has been used to examine negative ion chemical ionization (CI) reactions which can be used to differentiate isomeric compounds. A number of unusual anions formed in a conventional high pressure CI source were successfully produced in the FTMS with much lower reagent gas pressures. Exact mass measurements were used to establish the elemental compositions of these anions. Swept double resonance experiments were performed to elucidate ionic precursors in gas phase reactions. In some cases, the observed anions were found to arise from reactions with trace impurities that are present in the vacuum system or the CI reagent gas, such as water and oxygen.

Negative ion chemical ionization (NICI) processes can be very useful for the differentiation of isomeric polycyclic aromatic hydrocarbons (PAH) (1). These compounds arise from both natural and anthropogenic sources, and are commonly found in complex mixtures in the environment (2). The physical and biological properties of PAH are known to vary with molecular structure (3). Thus, it is often important to be able to distinguish isomeric PAH in complex mixtures. Conventional electron impact ionization mass spectrometry is often limited in its ability to identify PAH unambiguously because these compounds typically yield intense molecular ions with little or no additional fragmentation which would give insight into the identity of a particular isomer.

In the NICI technique (1), methane was used as a buffer gas to produce thermal electrons, which were preferentially captured by an analyte species to form anions. In the case of PAH, it was observed that the ionization behavior of these compounds under NICI conditions was related to the electron affinity (EA) of the compound. Compounds with EA values greater than about 0.5 eV captured electrons to form molecular anions, whereas PAH with lower EA values did not ionize. This difference in ionization behavior

0097-6156/87/0359-0175$06.00/0

under NICI conditions allowed a number of isomeric PAH to be distinguished readily. For example, the potent carcinogen benzo(a)pyrene (EA = 0.64 eV (4)) could be selectively detected and quantitatively determined in the presence of its relatively benign isomer, benzo(e)pyrene (EA = 0.32 eV (4)), without interference.

A number of other classes of aromatic compounds have also been studied to determine the generality of the observed 0.5 eV ionization threshold. For example, azaarenes (which have a nitrogen in the aromatic ring) were found to follow the same trend as the PAH, allowing many isomers of this class of compounds to be distinguished on the basis of relative electron affinity values. Thus far, two classes of compounds have been found to be exceptions to the observed trend in ionization, including condensed aromatic systems which contain a saturated five-membered ring and amino-substituted PAH (aromatic amines). In the first case, compounds such as fluorene were not predicted to ionize under NICI conditions because they are typically very difficult to electrochemically reduce (5) and would be expected to have very low electron affinity values (6). Although molecular anions were not observed for these types of compounds, anions corresponding to $(M + 14)^-$ were readily formed.

A similar type of behavior was observed in the case of aromatic amines. Amino-substituted PAH, such as amino-fluoranthenes, which would be predicted to have EA values greater than 0.5 eV, did yield molecular anions in a manner similar to non-substituted PAH. However, aromatic amines which have lower EA values were also observed to produce anions under NICI conditions, although not molecular anions. For example, in the NICI spectrum of 1-aminoanthracene, the base peak corresponds to $(M + 14)^-$, while for 2-aminoanthracene, the base peak corresponds to $(M + 12)^-$. The formation of two different anions from these isomeric compounds suggested that differentiation of positional isomers of aromatic amines might be possible using the NICI technique.

The quadrupole mass spectrometer which was used for these initial studies could provide only limited insight into the formation or structures of these unusual ions from fluorene and amino PAH. Experiments using substituted fluorenes, such as 1-, 2-, and 9-methylfluorene, 9-phenylfluorene, and carbazole, revealed that the $(M + 14)^-$ ion did not form if the C-9 position was blocked. Knowing that the $(M + 14)^-$ ion was formed by a reaction at the C-9 carbon, two possible structures could be drawn for this anion. One possibility would be 9-methylfluorene (structure I), which could arise from the addition of $CH_2$ to fluorene. Similar formation of adducts from methane buffer gas under NICI conditions has been reported (7, 8). Alternatively, fluorene could lose the two hydrogens at the 9-position and add oxygen to form 9-fluorenone (structure II)

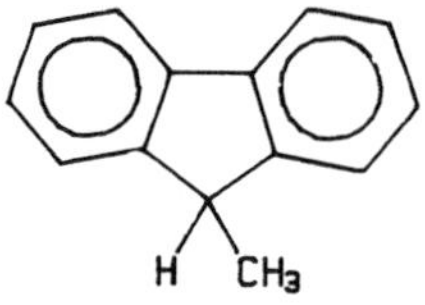

H CH3

I

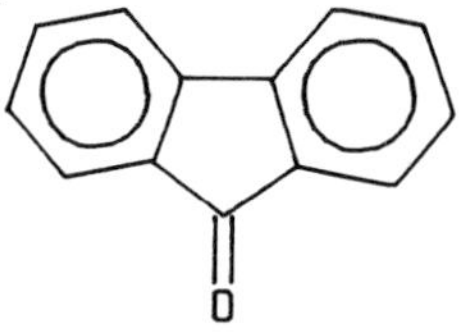

O

II

Fourier transform mass spectrometry (FTMS) is a powerful and versatile technique for investigating gas phase processes (9-12). Combined with ion trapping, the ability to perform double resonance and ion ejection experiments is helpful in probing gas phase reaction pathways. Furthermore, the high resolution and exact mass measurement capabilities of FTMS permit empirical molecular formulae of product ions to be easily established. Collisionally activated dissociation (MS/MS) reactions are often helpful in probing ion structures, as well. In this paper, FTMS has been used to examine the reaction pathways involved in a number of NICI reactions that can be used to differentiate isomeric compounds. Elucidation of these pathways might make it possible to devise additional reactions which would distinguish a wider variety of isomeric compounds using NICI processes. The conditions required to obtain NICI spectra on the Nicolet FTMS 2000 will be compared to those of more conventional high pressure CI sources. Finally, the experiment sequences, which were written to help elucidate the gas phase reaction pathways observed in these NICI studies, will be described.

## Experimental

Conventional high pressure NICI spectra were obtained using a Hewlett-Packard 5985B quadrupole GC/MS, as described previously (1). Methane was used as the CI reagent gas and was maintained in the source at 0.2-0.4 torr as measured through the direct inlet with a thermocouple gauge. A 200 eV electron beam was used to ionize the CI gas, and the entire source was maintained at a temperature of 200° C. Samples were introduced into the spectrometer via the gas chromatograph which was equipped with a 25 meter fused silica capillary column directly interfaced with the ion source. For all experiments, a column coated with bonded 5% methyl phenyl silicon stationary phase, (Quadrex, Inc.) was used and helium was employed as the carrier gas at a head pressure of 20 lbs. Molecular sieve/silica gel traps were used to remove water and impurities from the carrier gas.

The FTMS experiments were performed with a Nicolet Instruments, Inc. FTMS-2000 equipped with two differentially pumped 2 inch cubic cells and operated at a nominal magnetic field strength of 3 tesla. The differential pumping system is equipped with two 1,000 L/s oil diffusion pumps. A metal plate with a 2 mm diameter hole serves as a vacuum conductance limit between the cells and supports a pressure difference of approximately 1000:1. For low pressure CI experiments ($10^{-6}$ to $10^{-4}$ torr), the CI reagent gases were introduced into the source cell of the spectrometer through a batch inlet system equipped with 1 liter expansion volumes and molecular leaks. For higher pressure experiments (> $10^{-4}$ torr), the CI gas was introduced into the system by means of a pulsed valve arrangement. The reagent gases were ionized with electron energies typically at 70 eV, although energies as low as 10 eV were used. All samples were introduced into the spectrometer with a heated direct insertion probe, which was operated at temperatures between ambient and 250° C. Pressures in the FTMS cells were measured with ionization gauges. Unless mentioned otherwise, the FTMS cells were maintained at ambient temperature (about 30°C).

All standard compounds were obtained commercially and used as received. Labelled water, $D_2O$ (99.78 atom % D) and $H_2{}^{18}O$ (97.3 atom % $^{18}O$), were obtained from KOR Isotopes (Cambridge, MA) and MSD Isotopes (Montreal, Canada), respectively. Labelled oxygen, $^{18}O_2$ (52.6 atom % $^{18}O$), was also obtained from MSD Isotopes. Water samples were subjected to several freeze-pump-thaw cycles to remove volatile gases prior to introduction into the FTMS.

Several experiment sequences were written for the FTMS-2000 specifically for the NICI studies. One sequence allows ions to be formed in the source cell under relatively high static pressures (up to $10^{-4}$ torr) followed by transfer of the product ions into the analyzer cell for detection. An important feature of this experiment sequence is the use of a relatively high trapping potential (> 2.5 volts) in the source cell and a much lower trapping potential (< 1 volt) in the analyzer cell. This has the effect of greatly improving the trapping efficiency of the ions under the high pressure CI conditions, while allowing them to be detected in the analyzer under the influence of weaker electric fields for improved exact mass measurements.

For experiments requiring higher pressures of CI reagent gas, a second experiment sequence was written which incorporates the use of the pulsed valves supplied with the FTMS. In this sequence, the pulsed valves are opened for 6 msec, allowing a high pressure pulse of reagent gas to enter the source cell. A variable delay is introduced between the time the valves are opened and the time the electron beam is turned on. This enables the pressure at which the initial ionization occurs to be varied. Again, after ion formation in the source, the products may be transferred into the analyzer cell for detection if desired.

In order to help elucidate reaction pathways by which the product anions are formed in the NICI reactions, a Fortran program was written to perform a swept double resonance experiment. With this program, the intensity of a product ion is monitored while a suspected reactant ion is continuously ejected from the FTMS cell, prohibiting it from participating in the formation reaction. Ejection of a precursor ion results in a decrease in the measured intensity of the product ion. Experimentally this information is obtained by collecting a series of spectra in which the mass of the ejected ion is incremented by a specified amount. This program allows the user to specify both the mass limits and mass increment of the ion ejection event. The number of spectra collected at each ion ejection mass may also be determined by the user. Results are tabulated on a printer in units of absolute ion intensity for the product ion of interest versus the mass of the ejected ion. A graphic representation of the data is also generated on a plotter. Two versions of this program allow one to either eject ions formed during the electron beam or during the ion trapping time after the beam.

All experiment sequences and Fortran programs described above are available from the authors on 8 inch floppy disks. In addition, source codes and documentation for each program is available.

## Results and Discussion

Comparison of Chemical Ionization Conditions. A comparison of the conditions used in the conventional high pressure NICI experiment and the FTMS experiment are outlined in Table I. One difference between the two instruments is the reagent gas pressure used for chemical ionization. In the case of conventional CI experiments, the CI reagent is introduced at relatively high pressures (a few tenths of a torr) to generate a plasma of ions, electrons, neutrals and radicals in the ionization source (13). In the case of FTMS, however, CI reactions can be performed in several different manners. First, reagent gases may be introduced into the spectrometer at relatively low pressures ($10^{-6}$ to $10^{-8}$ torr), and a delay time (up to several seconds) inserted between reagent ion formation and detection of product ions to allow gas phase reactions to occur (14). Alternatively, pulsed valves (15, 16) may be used to introduce brief high pressure pulses of gas into the FTMS cell to approximate conventional CI conditions more closely and still maintain most of the high performance capabilities of the FTMS. Finally, with the differentially pumped dual cell design of the FTMS 2000 (17), a somewhat higher static pressure ($10^{-5}$ to $10^{-4}$ torr) of the CI reagent gas may be maintained in the source cell, while pressure in the analyzer cell is typically three orders of magnitude lower. The ions formed in the source cell under CI conditions can be transferred into the analyzer cell, allowing them to be detected at pressures which permit much higher performance of the FTMS, including high mass resolution and exact mass measurement. In this study, most spectra were obtained with a static pressure of $10^{-6}$ to $10^{-4}$ torr of reagent gas in the source cell, although pulsed valves were used to generate higher pressures in a few experiments. Reaction times of up to 30 seconds were used before ion detection in order to observe anion formation.

Table I. Comparison of NICI Conditions

| Parameter | quadrupole | FTMS |
|---|---|---|
| Pressure | 0.2 -0.4 torr | $10^{-6}$ - $10^{-4}$ torr |
| Electron Beam | continuous | < .001 - > 1 sec. |
| Detection | < msec | > msec |
| Temperature | ambient - 300°C | ambient - 250°C |

Another difference between conventional high pressure CI and the FTMS CI experiment is the duration of the electron beam event used to form primary ions and secondary electrons. In the conventional CI source, the electron beam is on continuously during the experiment. The FTMS, in contrast, uses a pulsed electron beam, and the duration of the electron beam event may be varied from less than a millisecond to over a second. In the NICI studies using the FTMS, the electron beam was typically left on for ten milliseconds. However, it was found that in some cases, which will be discussed later, it was necessary to use a longer beam time, up to 1 sec, in order to observe the product ions normally produced under the high pressure NICI conditions used on the quadrupole instrument.

Finally, the time required for the detection of ions also differs between the quadrupole instrument and the FTMS. In the quadrupole instrument, ions are typically detected in less than a millisecond after leaving the ion source, whereas the FTMS experiment normally requires tens of milliseconds to acquire a medium resolution spectrum. Therefore, it is conceivable that short-lived anions, which may be observed with the quadrupole instrument, may not be readily detected with the FTMS.

In order to assess whether the FTMS could be used to simulate conventional high pressure NICI reactions, a mixture of fluorene, fluoranthene, and benzo(a)pyrene was investigated. This mixture was introduced into the instrument using the heated solids probe and the resulting spectrum is shown in Figure 1. This spectrum was obtained with what might be called "typical" CI conditions for the FTMS 2000, with the source side of the cell maintained at a static pressure of $7 \times 10^{-6}$ torr of methane, a 10 msec electron beam, and a delay time of 30 s after ion formation to allow post-ionization reactions to occur. For all three compounds, the anions formed were identical to those observed using the quadrupole instrument, with molecular anions at m/z 202 and 252 produced from fluoranthene and benzo(a)pyrene, respectively, while an anion at m/z 180, corresponding to $(M + 14)^-$, was produced from fluorene.

In the studies outlined below, collisionally activated dissociation (MS/MS) experiments were attempted as a means of helping establish the identities of a number of anions. In every case tried, however, the parent anions appeared to undergo electron detachment with no daughter ion production observed.

<u>Formation of M + 14 Anions From Fluorene.</u> One of the first NICI experiments performed on the FTMS was the investigation of the peak at m/z 180 observed from fluorene, to determine whether the species formed was similar to 9-methylfluorene (I) or 9-fluorenone (II). Exact mass measurement of this anion yielded a value of 180.059, which compared favorably to that expected for the fluorenone structure of $C_{13}H_8O$ (mass 180.058) rather than the 9-methylfluorene structure of $C_{14}H_{12}$ (mass 180.094). To verify this conclusion, a mixture of fluorene and 9-methyl fluorene was introduced via the solids probe. The resulting NICI spectrum (18) is shown in Figure 2, in which two distinct ions are observed. One corresponds to the peak expected for 9-methyl fluorene and the other to 9-fluorenone, indicating that under NICI conditions, fluorene loses two hydrogens at the C-9 position and adds an oxygen atom to form a species similar to 9-fluorenone. It should be noted that in this particular spectrum, argon was used as the buffer gas, instead of methane, demonstrating that methane does not participate in the gas phase reaction. In addition, reactions using $^{13}CH_4$ and $CD_4$ as reagent gases also failed to indicate the participation of methane in the formation of the m/z 180 anion. Experiments using 9,9-$d_2$-fluorene (molecular weight 168) also yielded anions at m/z 180, which corresponds to (M + O - 2D), further supporting the conclusion that the reaction occurs at the C-9 position.

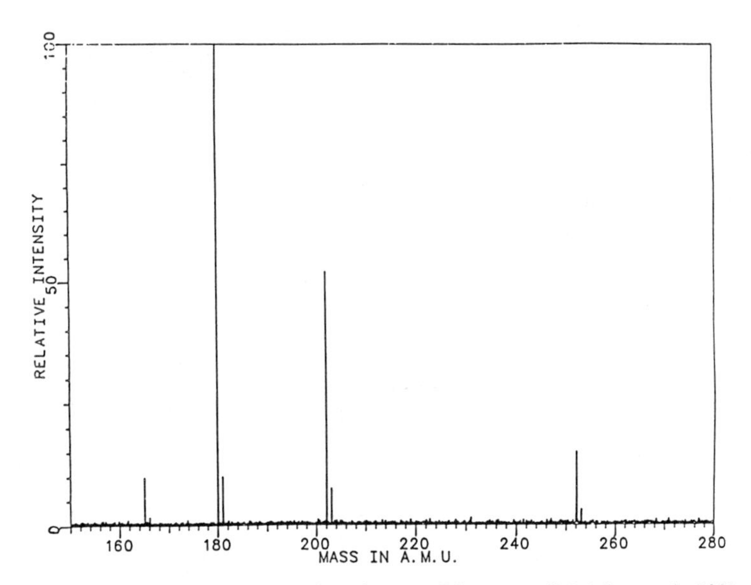


Figure 1. FTMS NICI spectrum of a mixture of fluorene [(M+14) at m/z 180], fluoranthene [$M^-$ at m/z 202)], and benzo(a)pyrene [($M^-$ at m/z 252)]. Methane was used as the reagent gas at a static pressure of 7 x $10^{-6}$ torr.

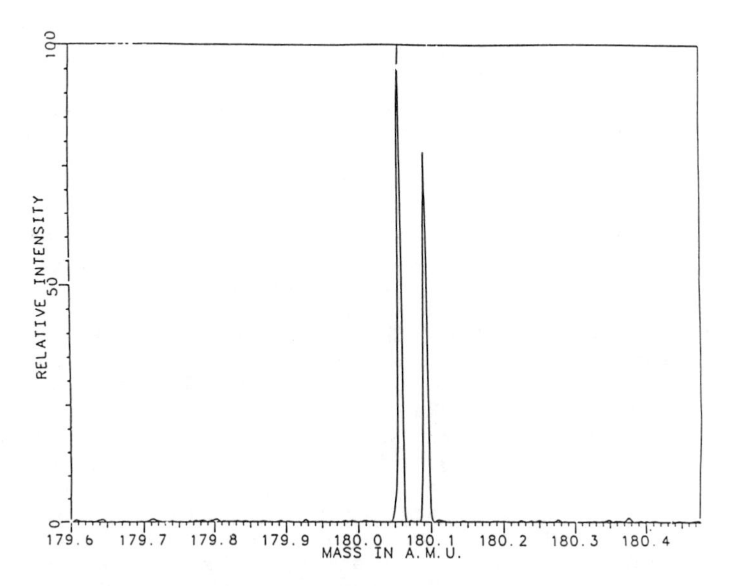


Figure 2. FTMS NICI spectrum of a mixture of fluorene and 9-methyl fluorene, using argon as the reagent gas. Observed anions at 180.058 and 180.094 correspond to $C_{13}H_8O^-$ and $C_{14}H_{12}O^-$, respectively. (Reproduced with permission from ref. 18. Copyright 1987 Wiley.)

Knowing that the anion at (M + 14) mass units exhibited oxygen incorporation, experiments were conducted with labelled reagent gases, including $H_2{}^{18}O$ and $^{18}O_2$, to determine the source of the oxygen. Initial experiments using these labelled reagents and the conditions typically used for the NICI reactions on the FTMS gave no indication that either water or molecular oxygen were reacting with fluorene in the gas phase. Increasing the ion trapping time prior to detection also did not yield any species in which labelled oxygen was incorporated. The species at $(M + 14)^-$ was still observed in both of these cases, but no ion at $(M + 16)^-$, corresponding to $(M - 2H + {}^{18}O)^-$, was observed.

When the electron beam time was increased to longer times (up to 500 msec), additional anions were observed. This is illustrated in Figure 3, which is the NICI spectrum obtained from a reaction of 9,9-$d_2$-fluorene (molecular weight 168) with oxygen-18 using an electron beam time of 200 msec. A significant anion is observed at m/z 166, due to loss of a deuterium atom from the C-9 position of 9,9-$d_2$-fluorene. Another anion observed at m/z 165 arises from the loss of a hydrogen from the fully protonated fluorene which was present as an impurity in the sample. Spectra obtained using a much shorter electron beam time did exhibit the characteristic anion at m/z 180 corresponding to (M + O - 2D) from the deuterated fluorene was observed, but the analogous anion with $^{18}O$ incorporated was not observed.

Additional anions were present in the NICI spectrum of 9,9-$d_2$-fluorene obtained with $^{18}O_2$ as the reagent gas and using a 200 msec beam time which did indicate the incorporation of $^{18}O$, as shown in Figure 3. Anions at m/z 181, 183, and 185 were found to correspond to the general formula of $(M + O - H)^-$, where m/z 181 arises from fluorene with $^{16}O$ incorporation, m/z 183 arises from $d_2$-fluorene with $^{16}O$, as well as fluorene and $^{18}O$, and m/z 185 is from $d_2$-fluorene and $^{18}O$. To help identify the precursor ions from which these products were formed, a swept double resonance experiment was performed. The results of this experiment indicated that $^{18}O^-$ was the ion which reacts with the 9,9-$d_2$-fluorene to produce the anion corresponding to $(M + {}^{18}O - H)$ at m/z 185, and that $^{16}O^-$ produces the anion at m/z 183. An interesting observation was the fact that this reaction involved the loss of a hydrogen from the ring system instead of the loss of a deuterium from the labelled C-9 carbon. This is in contrast to the reaction discussed earlier where the formation of the m/z 180 anion from fluorene was determined to be due to the addition of oxygen to the C-9 carbon with the loss of the two hydrogens at that position.

The reaction of fluorene with water was also studied using longer electron beam times, with the major anion produced being $(M - H)^-$. Swept double resonance was again used to study this reaction and the results are shown in Figure 4. In this experiment, perdeuterofluorene (molecular weight 176) was introduced via the solids probe and $H_2{}^{18}O$ as used as the NICI buffer gas. The anion at m/z 174, corresponding to $(M - D)^-$, was monitored as masses from 15 to 22 amu were ejected at 0.2 amu intervals. Four "peaks", corresponding to decreases in the intensity of the product ion at m/z 174 were observed in the resulting swept double resonance plot, at 17, 18, 19, and 20 amu,

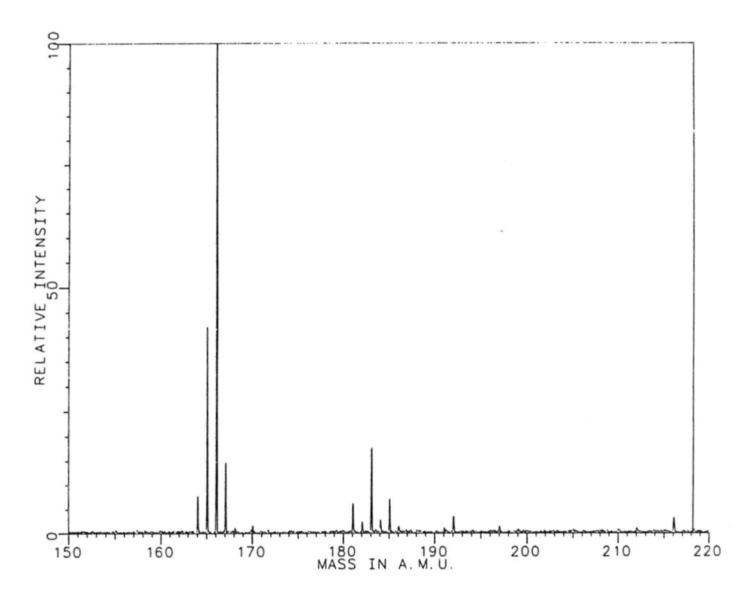


Figure 3. FTMS NICI spectrum of 9,9-$d_2$ fluorene using $^{18}O_2$ as a reagent gas ($1 \times 10^{-5}$ torr) and a 200 msec electron beam.

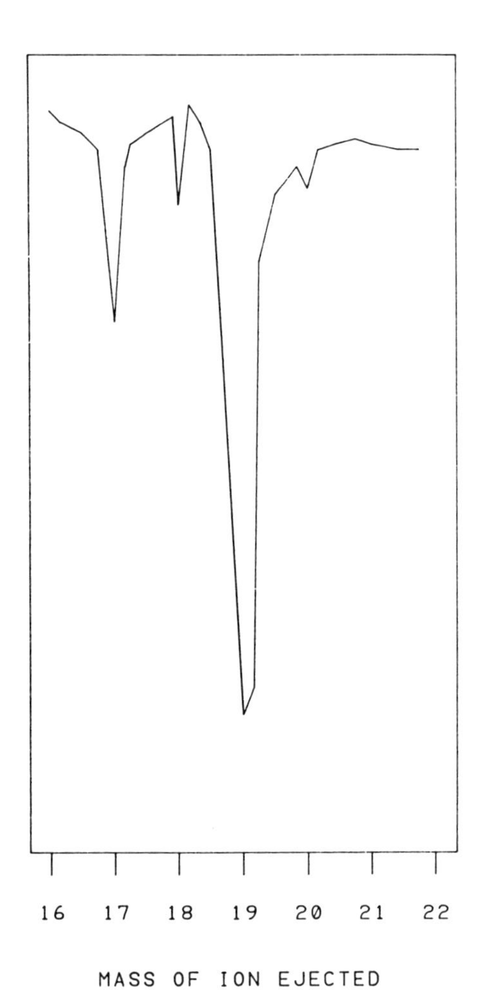


Figure 4. Swept double resonance experiment on perdeuterofluorene using $H_2{}^{18}O$ as a reagent gas. Mass 174 was monitored as masses from 15 to 21 amu were sequentially ejected at 0.2 amu intervals.

corresponding to masses $^{16}OH^-$, $^{16}OD^-$, $^{18}OH^-$, and $^{18}OD^-$, respectively. This experiment confirms that these four ions (which originate from the buffer gas, $H_2{}^{18}O$, and traces of $H_2O$, as well as from the H/D exchange reactions of these two species with the deuterated fluorene) react with neutral fluorene in the gas phase to produce the anion at m/z 174.

A summary of the observed reactions with fluorene is given in Table II. In none of the NICI experiments conducted with the FTMS to date has $^{18}O$ from either labelled water or oxygen ever been observed to be incorporated into fluorene to form the analogous anion, $(M + {}^{18}O - 2H)^-$ at m/z 182. Thus, the precursor for this reaction has not been identified. These experiments have involved changing a variety of experimental conditions, including cell temperature, pressure, trapping potential, reaction delay time, electron beam time, and electron energy. Recently, small quantitites of $^{18}O_2$ were doped into the methane reagent gas used in the high pressure source on the quadrupole instrument and an anion at m/z 182 was observed, indicative of incorporation of $^{18}O$, although the extent of $^{18}O$ incorporation was relatively low. It is possible that the formation of the M + 14 anion from fluorene may proceed by different pathways in the two instruments.

Reactions of 1- and 2-aminoanthracene. As pointed out earlier, high pressure NICI spectra obtained with the quadrupole mass spectrometer yielded a possible means of differentiating positional isomers of aminoanthracene, with 1-aminoanthracene producing a base peak at (M + 14) mass units and 2-aminoanthracene, at (M + 12) mass units. Initial FTMS studies of NICI spectra of 2-aminoanthracene obtained with the cell at ambient temperature yielded a base peak at m/z 192, corresponding to $(M - H)^-$, as well as anions at m/z 193 ($M^-$) and m/z 204. Using the conventional high pressure CI source on the quadrupole mass spectrometer, which was operated at 200° C, the base peak was m/z 205, with anions also observed at m/z 204 (about 60% relative abundance) and m/z 192 (about 50% relative abundance). A study was conducted on the quadrupole mass spectrometer to investigate the effects of ion source temperature on the product ions observed under negative ion CI conditions for 2-aminoanthracene and the results are shown in Figure 5. A marked increase in the formation of molecular anions was observed at lower temperatures. This observation is consistent with other reports of enhanced molecular anion lifetimes at low source temperatures for other types of compounds under negative ion CI conditions, such as azulene (19). At higher source temperatures, the abundance of the molecular anion decreased as the intensity of $(M + 12)^-$ increased. This same temperature effect was observed in the case of $(M + 14)^-$ at m/z 207 for 1-aminoanthracene as well. Based on this information, the temperature of the FTMS cell was increased to 200° C to more effectively reproduce the conditions of ion formation in the conventional high pressure CI source.

The FTMS negative ion CI spectra of 1-aminoanthracene exhibited anions at m/z 192 (base peak), 207, and 208 when the cell was maintained at 200° C and methane was used as the reagent gas at pressures of about $10^{-6}$ torr or higher. As pointed out above, as the cell temperature was decreased, the intensity of the anion at

Table II

SUMMARY OF NICI REACTIONS OF FLUORENE CHARACTERIZED BY FTMS

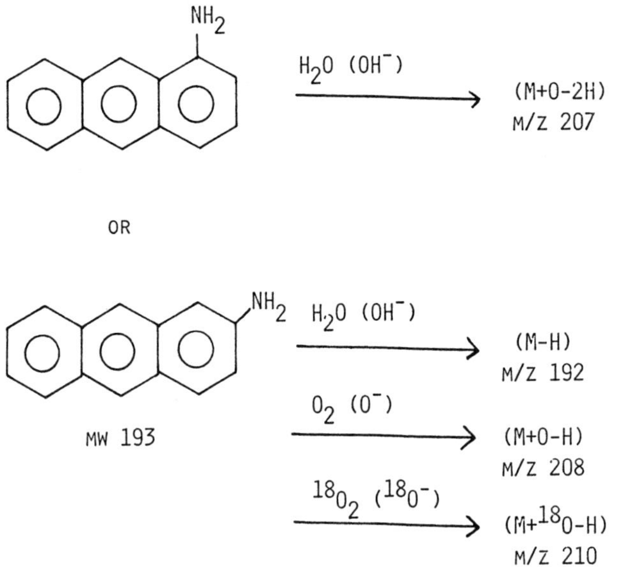

$NH_2$

$H_2O$ ($OH^-$) ⟶ (M+O-2H) M/Z 207

OR

$NH_2$

$H_2O$ ($OH^-$) ⟶ (M-H) M/Z 192

$O_2$ ($O^-$) ⟶ (M+O-H) M/Z 208

$^{18}O_2$ ($^{18}O^-$) ⟶ (M+$^{18}$O-H) M/Z 210

MW 193

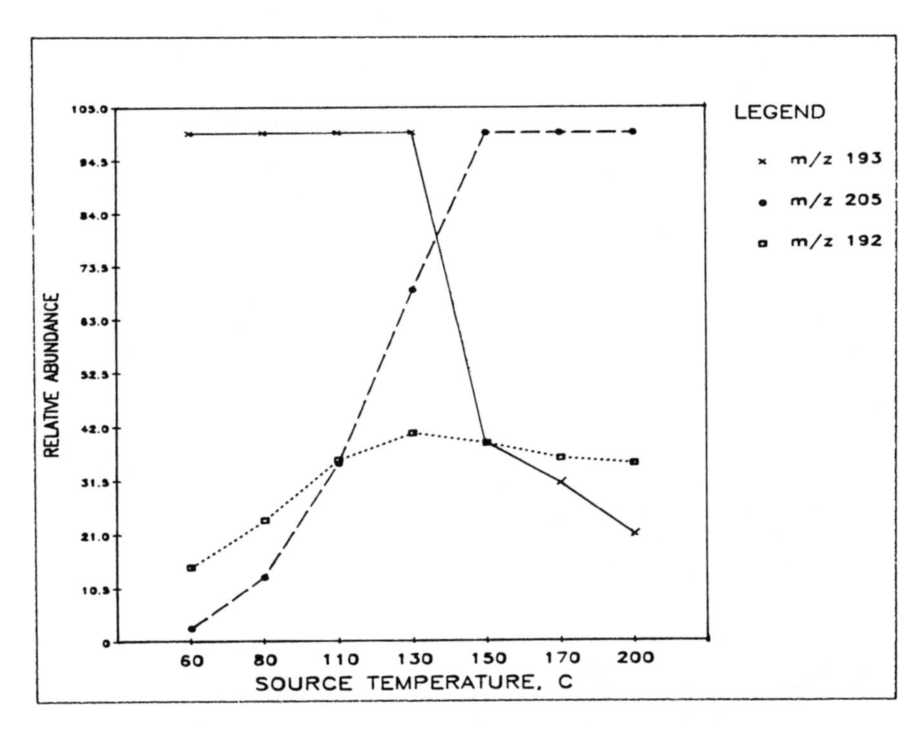


Figure 5. Effect of source temperature on anion formation from 2-aminoanthracene using conventional high pressure CI.

m/z 192 increased relative to that of the m/z 207 and 208 anions. Swept double resonance experiments indicated that both the m/z 192 and m/z 207 anions were formed by reaction of $OH^-$ with 1-aminoanthracene. Exact mass measurements suggested that these ions corresponded to $C_{14}H_{10}N$ (M - H) and $C_{14}H_9NO$ (M + O - 2H), respectively. Using swept double resonance, the anion at m/z 208 was found to be linked to the reaction of $O^-$ with neutral 1-aminoanthracene. The measured mass of the m/z 208 corresponded to $C_{14}H_{10}NO$ (M + O - H). Additionally, the analogous anion (M + $^{18}O$ - H) at m/z 210 was observed when $^{18}O_2$ was introduced as the reagent gas. The reactions observed for 1-aminoanthracene are sumarized in Table III.

For 2-aminoanthracene, m/z 192 was also observed to be the base peak at a cell temperature of 200° C, using methane as the reagent gas. In addition, anions at m/z 204 and 205 were observed (approximately 50% and 11% relative abundance, respectively). When $H_2O$ was used as the CI reagent, anions at m/z 192 (M - H), 207, and 208 were observed. As in the case of 1-aminoanthracene, double resonance again showed that the (M - H) anion was produced from the reaction of $OH^-$ with the parent molecule. The anion at m/z 207 was found to be linked to $OH^-$ as well, and exact mass measurement revealed that this ion corresponded to an empirical formula of $C_{14}H_9ON$, or (M + O - 2H). The formation of the m/z 208 anion was found to arise from the reaction of $O^-$ with 2-aminoanthracene. According to exact mass measurements, this anion corresponded to $C_{14}H_{10}NO$ or (M + O - H). The participation of oxygen in this reaction was verified using $^{18}O_2$ as the CI reagent gas, which produced an anion at m/z 210, corresponding to (M + $^{18}O$ - H).

Exact mass measurement of the m/z 205 anion indicated that this species corresponded to $C_{14}H_7NO$, or (M + O - 4H). Double resonance experiments did not identify a link to an oxygen-containing precursor, including $OH^-$, $O^-$, or $O_2^-$. The measured mass of the m/z 204 anion indicated an empirical formula of $C_{14}H_8N_2$, or (M + N - 3H). The m/z 205 and 204 anions were only observed when methane was used as the CI reagent, and not when either oxygen or water were used. Reactions with $^{13}CH_4$ as the reagent gas failed to show any participation of methane in the formation of either of these anions.

Additional experiments were conducted with the high pressure CI source on the quadrupole mass spectrometer using small amounts of $O_2$, $^{18}O_2$, and $H_2O$ doped into the methane reagent gas to ascertain whether oxygen was incorporated into the anions observed from 2-aminoanthracene. When a mixture of $O_2$ and $CH_4$ was used as the NICI reagent gas (producing $OH^-$ and $O_2^-$), the abundances of m/z 192, 205 and 207 were observed to increase, suggesting that oxygen was responsible for the formation of these ions in the high pressure CI source. When this experiment was repeated by adding $^{18}O_2$ to the methane reagent gas, an anion at m/z 209 was observed, further indicating that oxygen was participating in the formation of the m/z 207 anion. An increase in the 207 anion was also observed in this spectrum. However, the low mass resolution of the quadrupole mass spectrometer did not permit the determination of whether this increase was due to the incorporation of $^{18}O$ into the

Table III

SUMMARY OF NICI REACTIONS 1- AND 2-AMINOANTHRACENE CHARACTERIZED BY FTMS

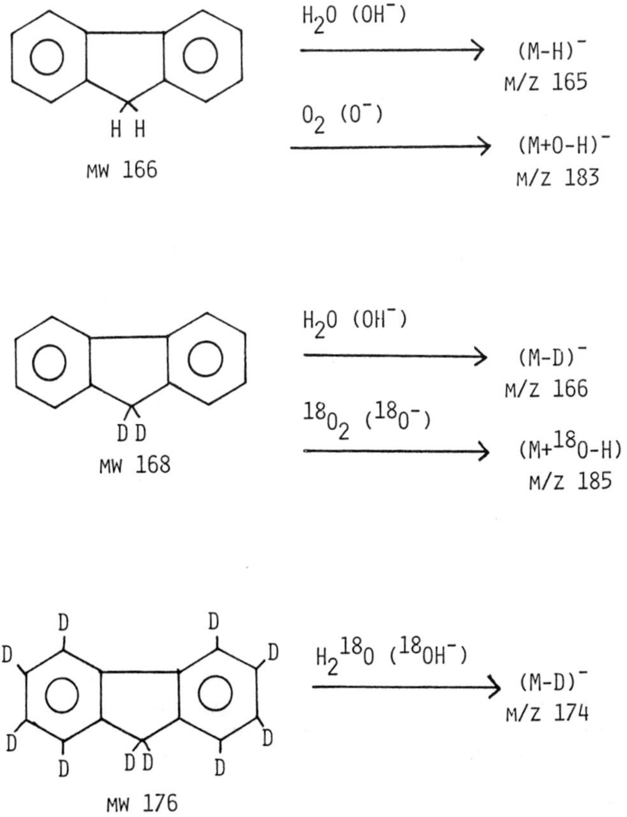

H H

MW 166

$H_2O$ ($OH^-$) → $(M-H)^-$ M/Z 165

$O_2$ ($O^-$) → $(M+O-H)^-$ M/Z 183

D D

MW 168

$H_2O$ ($OH^-$) → $(M-D)^-$ M/Z 166

$^{18}O_2$ ($^{18}O^-$) → $(M+^{18}O-H)$ M/Z 185

D D D D D D D D D D

MW 176

$H_2{}^{18}O$ ($^{18}OH^-$) → $(M-D)^-$ M/Z 174

m/z 205 anion or if it was from the original m/z 207 anion. When a mixture of water and methane was used as the NICI reagent gas, producing $OH^-$, only $(M-H)^-$ was enhanced. The FTMS studies indicated that the 207 anion was linked to $OH^-$. In these high pressure experiments, it is possible that the formation of this anion may proceed in a stepwise fashion and the presence of an additional reactant, such as $O_2$, may be required.

## Conclusions

Using the dual cell FTMS, it has been shown that a number of unusual ions formed in the high pressure CI source result from reaction with trace impurities present in the vacuum system or from the CI reagent gas, including water and oxygen. In some cases, these impurities are responsible for the production of anions which can be used to differentiate isomeric structures of molecules. This work suggests that doping CI reagent gases with controlled amounts of these impurities might enhance the reproducibility of these reactions so they could be used as an analytical method for isomeric differentiation.

FTMS has been shown to be a powerful tool for studying NICI reactions. In many cases, the observed NICI spectra were very complex, involving a number of reactions. The ability to form the anions under high pressure conditions and detect them at low pressures using the dual cell configuration, combined with several specialized experimental sequences, played a major role in unravelling these reactions. The results of these studies have shown that for simple electron capture reactions in which molecular anions are formed, the ionization in the lower pressure FTMS source is similar to that in the high pressure CI source on the quadrupole mass spectrometer. In cases where reactions other than simple electron capture occur, the observed anions, or possibly even the pathways by which they are formed, sometimes differ between the two types of sources. In order to minimize these differences, a high pressure CI source for the FTMS 2000 is being designed and constructed in our laboratory. This source should allow a much better means of studying chemical ionization reactions observed in conventional high pressure CI sources using FTMS.

## Acknowledgments

The authors thank Dr. Elizabeth Stemmler for obtaining the quadrupole data. Research sponsored by the Office of Health and Environmental Research, U.S. Department of Energy under contract DE-AC05-84OR21400 with Martin Marietta Energy Systems, Inc.

## Literature Cited

1. M. V. Buchanan and G. Olerich, Org. Mass Spectrom. **19**, 486 (1984).
2. R. E. Laflamme and R. A. Hites, Geochim. Cosmochim. Acta (1978), **42**, 289.
3. M. L. Lee, M. V. Novotny and K. D. Bartle, "Analytical Chemistry of Polycyclic Aromatic Compounds", Academic Press, New York, 1981.

4. J. M. Younkin, L. J. Smith, and R. N. Compton, Theor. Chim. Acta (Berl.) **41**, 157 (1976).
5. I. Bergman, Trans. Faraday Soc. **50**, 829 (1954).
6. A. Streitwieser, Jr., "Molecular Orbital Theory for Organic Chemists," pp.175-185, John Wiley and Sons, New York, (1961).
7. D. Stockl and H. Budzikiewicz, Org. Mass Spectrom. **17**, 376 (1982).
8. G. W. Dillow and I. K. Gregor, Org. Mass Spectrom. **21**, 386 (1986).
9. A. G. Marshall, M. B. Comisarow, and G. J. Parisod, J. Chem. Phys. **71**, 4434 (1979).
10. A. G. Marshall, Acc. Chem. Res. (1985) **18**, 316.
11. M. L. Gross and D. L. Rempel, Science, (1984), **226**, 261.
12. C. L. Wilkins and M. L. Gross, Anal. Chem., (1981), **53**, 1661A.
13. A. G. Harrison, "Chemical Ionization Mass Spectrometry", pp.75-80, CRC Press, Inc., Boca Raton, FL, (1983).
14. S. Ghaderi, P. S. Kulkarni, E. B. Ledford, Jr., C. L. Wilkins, and M. L. Gross, Anal. Chem. **53**, 428 (1981).
15. T. J. Carlin and B. S. Freiser, Anal. Chem. **55**, 571 (1983).
16. D. A. McCrery, T. M. Sack and M. L. Gross, Spectrosc. Int. J. **3**, 57 (1984).
17. R. B. Cody, J. A. Kinsinger, Sahba Ghaderi, I. J. Amster, F. W. McLafferty, and C. E. Brown, Anal. Chim. Acta, **178**, 43 (1985).
18. M. V. Buchanan, I. B. Rubin, M. B. Wise, and G. L. Glish, Biomed. and Environ. Mass Spectrom., **14**, 395 (1987).
19. E. P. Grimsrud, S. Chowdhury, and P. Kebarle, J. Chem. Phys. (1985) 83, 3983.

# Glossary

**A/D Converter:** Analog to digital converter; changes a variable electrical signal into a digital representation that can be processed by a computer.

**Aliasing:** Artificial shift of a high frequency signal to a lower frequency (fold-back) due to sampling the signal at a rate of less than two points per cycle; see Nyquist frequency.

**Analog Signal:** A continuously variable signal.

**Analyzer Cell:** General term used for the lower pressure trapped ion cell in dual cell FTMS instruments.

**Apodization:** Weighting function applied to a truncated time-domain signal in order to improve the peak shape of the frequency domain spectrum by reducing the side-lobes.

**Array Processor:** Computer circuit designed for performing fast numeric computations on large arrays of data.

**Bandwidth:** Frequency range over which a signal is attenuated by less than 3 dB.

**Base Peak:** Most intense peak in a mass spectrum.

**Baseline Correction:** Weighting scheme used to achieve a flat baseline in Fourier transformed signals.

**Beam Time:** The length of time that the electron beam is on during an experiment cycle.

**Broadband Excitation:** Irradiation of ions with a wide range of radio frequencies (typically in the range of kHz to MHz) to bring them into coherent motion in order to detect a wide mass range.

**CAD:** Collisionally Activated Dissociation; dissociation of a ion caused by selected excitation of that ion and subsequent collision with a target neutral species.

**CI:** Chemical Ionization.

**CID:** Collision Induced Dissociation (sometimes used interchangeably with CAD).

**Coherent Motion:** In-phase movement of ions within a trapped ion cell.

0097-6156/87/0359-0192$06.00/0

**Collision Energy:** Energy transferred to an ion upon collision with a gas or surface.

**Collision Gas:** A target gas (typically argon) used for collisionally activated dissociation of ions.

**Conductance Limit:** Orifice in a metal plate separating the two cells in a dual cell FTMS instrument; allows differential pumping of the two cells.

**Co-processor:** Additional circuit added to a computer to increase the speed of calculations.

**Cross Section:** Measure of the probability of an event taking place.

**Cyclotron Equation:** Inverse relationship between ionic mass (m) and cyclotron frequency ($\omega$), that is, $\omega = KqB/m$, where K is a constant, q is ionic charge, B is magnetic field strength.

**Cyclotron Frequency:** The angular frequency of the orbital motion of an ion in a constant magnetic field; ions of different masses have unique cyclotron frequencies.

**Cyclotron Motion:** Orbital motion of an ion or charged particle in a magnetic field.

**Dalton:** An atomic mass unit based on $^{12}C$ = 12 daltons.

**Damping:** Loss of coherent ion cyclotron motion primarily due to collisions with neutral molecules; results in loss of signal.

**Daughter Ion:** Product ion from the reaction or dissociation of a parent ion.

**Delay:** Length of time between two sequential events in an FTMS experiment sequence.

**Differential Pumping:** The use of two or more independent pumping systems to create different pressure regions within a vacuum system.

**Direct Mode:** Direct digitization of a broadband signal transient.

**Double Resonance:** Use of appropriate radiofrequency pulses to eject selected ion(s) from the trapped ion cell.

**Dual Cell FTMS:** FTMS instrument configuration that uses two differentially pumped trapped ion cells.

**Dynamic Range:** The difference between the maximum and minimum number of ions that can be detected without signal distortion.

EI: Electron ionization.

EIEIO: Electron Impact Excitation of Ions from Organics; uses low energy electrons to collisionally activate and dissociate ions.

**Electron Affinity:** Energy required to remove an electron from an anion in its ground state.

eV: Electron volt.

**Excitation Amplifier:** Electronic circuit designed to increase the amplitude of the rf signal generated by the frequency synthesizer.

**Excitation Level:** Magnitude of the rf voltage applied to the FTMS cell to induce coherent motion of the ions.

**Excite Signal:** Radio frequency pulse applied to the transmit plates of the FTMS cell to induce coherent motion of the ions.

**Experiment Sequence:** Series of linked events controlled by the computer which includes ion formation, manipulation, and detection, among others.

**External Source:** Ionization region which is physically separated from the FTMS cell and usually independently pumped.

**FAB:** Fast Atom Bombardment.

**Fast Fourier Transform:** An algorithm which calculates the Fourier transform of a digitized signal efficiently with respect to time and computer memory space.

**Fixed Frequency Excitation:** Selective excitation of a single mass by applying a fixed frequency to the transmit plates of the FTMS cell.

**Forward Fourier Transform:** Mathematical transformation of a signal from the time domain to the frequency domain.

**Fourier Transform:** Mathematical method of determining the frequency components (amplitude and phase of real and imaginary components) of a complex time domain signal.

**Frequency Domain Spectrum:** Graphic representation of spectral information in the form of amplitude versus frequency.

**Frequency Synthesizer:** Specialized circuit (usually computer controlled) which generates fixed or swept radiofrequency signals.

**Frequency Chirp:** Fast broadband radiofrequency sweep used to induce coherent ion motion in the FTMS cell.

**FTICR:** Fourier Transform Ion Cyclotron Resonance (used interchangeably with FTMS).

**FTMS:** Fourier Transform Mass Spectrometry (used interchangeably with FTICR).

**GC/MS:** Gas Chromatography/Mass Spectrometry; technique which utilizes a gas chromatograph interfaced to mass spectrometer, especially for the on-line analysis of mixtures.

**Heterodyne:** Process of mixing two frequencies (signal and reference) together to produce a lower frequency signal which can be digitized at a slower rate than the original signal; commonly used in FTMS to attain ultrahigh mass resolution spectra.

**ICR:** Ion Cyclotron Resonance Spectrometry.

**Image Current:** Time-varying electrical current induced in the receiver plates of the FTMS cell by coherent ion motion.

**Inverse Fourier Transform:** Mathematical transformation from the frequency domain to the time domain.

**Ion Ejection:** Excitation of an ion to a sufficiently large orbital to cause neutralization on the cell plates; used to remove ions selectively from the FTMS cell.

**Ion Cyclotron Resonance:** Type of mass spectrometry in which ion masses are determined on the basis of their unique orbital frequencies in a magnetic field.

**Ion Partitioning:** Distribution of the ion population between the source and analyzer cells in a dual cell FTMS instrument.

**Isobaric Ions:** Ions of the same nominal mass but having different elemental composition.

**LC/MS:** Combined liquid chromatography/mass spectrometry; technique which utilizes a liquid chromatograph interfaced to a mass spectrometer for the on-line analysis of mixtures.

**LDI:** Laser Desorption Ionization.

**m/z:** Mass to charge ratio.

**Magnitude Transform:** Absolute value representation of points in a frequency domain spectrum.

**Mixed Mode Detection:** Excitation of ions with a broadband radiofrequency sweep and heterodyne detection of a narrow frequency (mass) range.

**Molecular Ion:** Ion produced when a molecule gains or loses an electron.

**MPI:** Multiphoton Ionization; ionization of a molecule by two or more sequential photon absorption processes.

**MS I:** First stage of mass selection in an MS/MS experiment.

**MS II:** Second stage of mass selection in an MS/MS experiment.

**MS/MS:** Mass spectrometry/mass spectrometry; technique utilizing two stages of mass selection.

**$MS^n$:** Abbreviation for multiple stage mass spectrometry/mass spectrometry; a technique utilizing (n) stages of mass selection.

**Multiplex Advantage:** Time advantage gained by obtaining all spectral information simultaneously.

**Narrowband Excitation:** Application of a narrow range of radio frequencies to the transmit plates of the FTMS cell for the purpose of exciting a small window of ion masses.

**NICI:** Negative Ion Chemical Ionization; NCI is sometimes used interchangeably.

**Non-resonant Ion:** An ion whose cyclotron frequency is different than the excitation frequency applied to the FTMS cell.

**Nyquist Frequency:** Highest frequency which can be accurately represented at a specified sampling rate; according to sampling theory, the Nyquist frequency corresponds to one half the sampling frequency.

**Parent Ion:** Ion which dissociates or reacts to form daughter ions and neutrals.

**Particle Desorption:** Use of a high energy particle to desorb and ionize molecules from a surface.

**Photodissociation:** Fragmentation of an ion using photons.

**Pressure Broadening:** Increase in spectral linewidths due to collisional damping of the time domain signal.

**Proton Affinity:** Negative of the enthalpy change for the addition of a proton to a neutral species.

**Pulsed Valves:** Solenoid driven valves commonly used with FTMS to admit a brief burst of gas into the vacuum chamber; typically used to avoid operation of the instrument under high static pressure conditions in order to maintain high performance capabilities.

**Q-FTMS:** See Tandem quadrupole/FTMS.

**Quench Pulse**: A direct current potential applied to the trapping plates of the FTMS cell to remove all ions prior to the start of an experiment sequence.

**Reagentless CI**: Ion-molecule reactions between trapped fragment ions and neutral molecules (M) which result in the formation of ions from the latter.

**Resolution**: The degree of separation between two adjacent peaks in a spectrum; typically defined as the mass of the peak divided by the width of the peak at half its maximum height.

**Resonant Ion**: An ion in the FTMS cell whose cyclotron frequency is equal to the frequency of the rf excitation signal applied to the cell.

**rf Burst**: Application of a short pulse of rf voltage to the FTMS cell; may be used for ion detection or ejection.

**rf**: Radio frequency.

**S/N**: Signal-to-noise ratio.

**Sector MS**: A mass spectrometer that uses one or more electric and/or magnetic dispersion units to effect spatial mass separation.

**Self-CI**: Ion-molecule reactions between trapped daughter ions and neutral parent molecules (M) which result in the formation of ions from the latter, generating a prominent $(M + H)^+$ peak.

**SID**: Surface Induced Dissociation.

**Signal Amplifier**: An electronic circuit which increases the amplitude of the image current signal induced in the receiver plates of the FTMS cell.

**Signal Averaging**: Addition of successive spectral scans for improved signal-to-noise by reducing the level of random noise; applied to transients prior to Fourier transformation.

**SIMS**: Secondary Ion Mass Spectrometry.

**Single Resonance**: Experimental sequence using ion formation and detection events without ion ejection steps.

**Single Cell**: FTMS instrument design which uses one trapped ion cell.

**Source Cell**: Designation of the trapped ion cell located in the higher pressure region of a dual cell FTMS instrument.

**Space Charge**: Coulombic interaction of ions and/or electrons in an FTMS cell, which can lead to signal distortion.

**Surface Induced Dissociation:** Fragmentation of an ion induced by collision with a surface.

**SWIFT:** Stored Waveform Inverse Fourier Transform.

**T:** Tesla; unit of magnetic field strength.

**Tandem MS:** Two or more successive stages of mass spectrometry.

**Tandem Quadrupole FTMS:** Use of a quadrupole mass filter for introducing ions into the FTMS for subsequent analysis; used to circumvent pressure limitations.

**Thermal Energy:** Energy of an electron, ion, or molecule at ambient temperature.

**Time of Flight:** TOF; A type of mass analyzer which attains mass separation based on the length of time required for an ion of a known kinetic energy to traverse a specific distance.

**Time Domain Signal:** Complex sinusoidal ion image current which is digitized and transformed to yield the frequency domain mass spectrum.

**Transient:** Time domain signal equivalent to the ion image current.

**Transmitter Plates:** Plates in the FTMS cell which are used to couple radio frequency excitation to ions.

**Trapped Ion Cell:** An electrostatic-magnetic ion storage device.

**Trapping Efficiency:** A measure of the ability of a trapped ion cell to store ions.

**Trapping Potential:** The magnitude of the voltage applied to the trapping plates of the FTMS cell to reduce ion motion parallel to the magnetic field.

**Trapping Frequency:** The frequency of ion motion in the z-axis of the FTMS cell (parallel to the magnetic field); this frequency is much lower than the cyclotron frequency.

**Z-Mode Excitation:** Excitation of ions at their trapping frequencies in a direction parallel to the magnetic field.

**Zero Fill:** Addition of a block of zero points to a time domain spectrum prior to transformation of the data; often improves resolution when the number of data points is limited.

RECEIVED August 26, 1987

# INDEXES

# Author Index



# Affiliation Index



# Subject Index

*Production and indexing by Linda R. Ross*
*Jacket design by Carla L. Clemens*

*Elements typeset by Hot Type Ltd., Washington, DC*
*Printed and bound by Maple Press, York, PA*